THE VISAGE: UNMASKED

SIX STRANGERS, SIX BATTLES AND SIX DIFFERENT VICTORIES

Saga of Six Teenagers
Mentored by Yashi Shukla and Anjana Anand

Editor: Anoushka Ray (Teenager)

INDIA • SINGAPORE • MALAYSIA

Notion Press

Old No. 38, New No. 6
McNichols Road, Chetpet
Chennai - 600 031

First Published by Notion Press 2019
Copyright © Yashi Shukla 2019
All Rights Reserved.

ISBN 978-1-64678-682-4

This book has been published with all efforts taken to make the material error-free after the consent of the author. However, the author and the publisher do not assume and hereby disclaim any liability to any party for any loss, damage, or disruption caused by errors or omissions, whether such errors or omissions result from negligence, accident, or any other cause.

While every effort has been made to avoid any mistake or omission, this publication is being sold on the condition and understanding that neither the author nor the publishers or printers would be liable in any manner to any person by reason of any mistake or omission in this publication or for any action taken or omitted to be taken or advice rendered or accepted on the basis of this work. For any defect in printing or binding the publishers will be liable only to replace the defective copy by another copy of this work then available.

Contents

Foreword 5
Preface 7
Acknowledgements 9

1. Anger 11
2. Denial 73
3. Guilt 101
4. Fear 139
5. Reconstruction 179
6. Depression 199

Mentors 243
Co-Authors 247

Foreword

I am delighted to write this foreword, not only because I know of the good work Anjana and Yashi are doing as counselors, but also because the topic is of huge interest to not just me but many other parents too. The book aims to let us into what is going on in a teen's head. We often tend to blame it on the teen hormones and fail to see the real issues and emotions. 'The Visage: Unmasked" seeks to keep before you the real issues faced by a teen.

We as parents want the best for our children. We often dream of them attending the best of the universities, grabbing a coveted job opportunity and as a result, enjoy a great lifestyle. We want to raise winners. To be fair to ourselves, we do it with the best of the intentions. We want life to offer them with opportunities that we did not have access to during our days. So we invest in them and do everything within our capacity to help them get into the 'university & programs of choice'.

Whose choice will this be? In our focus to raise winners and successful individuals, are we missing out on something very important? Are we forgetting to stop and to listen to who matter the most? In the entire rush, are we failing to acknowledge that we are talking about Gen

Z – who is eager to break away from stereotypes, willing to experiment and take time to shape their identity.

This book is a much needed reminder and an eye-opener of sorts. Yashi and Anjana, bring our attention to the most important parenting topic – mental & emotional well being of our children. These are stories written by teens exposing their vulnerable side. In my opinion it cannot become more honest and real than this.

Reading this book, you will understand how to create that safe space where your teens are comfortable to open up and discuss. You will also acknowledge how 'conversations with an open mind' can help you and your child not just bond, but also help create a beautiful tomorrow for the child.

Our biggest want in life is a beautiful tomorrow for our loved ones. Let us do the right things then.

Best Wishes!

Neela Kaushik
Founder, GurgaonMoms

Preface

As career counsellors, our purpose is to mentor and facilitate students achieve their optimum potential. Most of our interactions are with teenagers and the biggest challenge we encounter is that these teens have already given up on their dreams; neither do they want to dream. Their fear, anxiety and emotional suffering are the biggest hurdles in realizing their own potential.

We have been very successful in giving a sense of comfort to all our students and as a result, are able to connect with these teens on a deeper level. Out of this emerged our hypothesis *"every teen is troubled."* A theory, that has been validated by our teenaged children. That was also the point we realised that regardless of how successful and "arrived" we may be, in the race to ensure that our children become super successful, we rarely give importance to their mental and emotional wellbeing; and this in our opinion was the often ignored, extremely important subject to talk about.

This book is for parents of every teen. Remember, **the need of the hour is *"conversation"*.** We must create enough emotional comfort spaces for our children... a place where no-one is judging them and they are free to share what they truly feel. It may be the biggest disappointment for every parent's dreams; it may shatter their own ideals of what constitutes success but in the end it is worth it because we have to believe that Nothing, Nothing at all, can ever be more important than our kids' emotional and mental well-being.

Let us come together and stand by our kids in all times. "The Visage: Unmasked" is our first attempt to bring our kids and their experiences together to talk about the most pertinent issue of our children's lives" Their Emotional and Mental well Being!!!

"Only Happy children can grow up to be Successful Individuals"

Yashi Shukla
Anjana Anand

Acknowledgements

Thank you Diya, Shivi and Shubham for letting us peep into your lives and bringing Yashi and I together. Your dreams, your aspirations and your challenges inspired FYCGlobal where each day, the energy of our children motivates us to surge ahead.

Thank you Sandeep Shukla and Vivek Anand for being the pillars of support all along.

Thank you Darshini Shah, Ira Bhattacharjee, Nandini Shukla, Navya Sheoran, Sadhika Anand and Samira Bhayana for putting your thoughts into words and penning your stories as co-authors.

Thank you Anoushka Ray for putting those countless nights into editing those drafts and re-drafts.

Thank you to all the contestants for making the judges' task difficult with your quality pieces of writing.

Thank you to the Almighty for guiding us.

Anger

– Ira Bhattacharjee

Anger-I

Ink blue skies explode into dazzling white lightning — the sound of thunder surrounds her as she realises that she's too large for this world, too large to be contained. There's a rage growing inside her, holding her back. She meets another, a dreamer like her — one with a bitter smile. They learn to dream together, learn to hold on to each other when dreams turn to nightmares, exchanging icy words and stolen kisses.

Anger meets Hate; a song and dance on the cruelty of fate. she is Anger, and she is Hate, twirling together in this bitter hate, pot calls the kettle black and they both go up in flames.

I've been told to write down my experiences on paper in hopes that I'd remember them forever. The who and the what that told me to do so doesn't matter, at least not yet. So here I am, revisiting memories of nearly a decade ago.

Of all my memories, the most significant one is of my meeting with Kyra. Kyra wasn't normal, at least not to me. No matter how anyone chose to explain it or call her some name or the other, assign her a non-existent disease or look upon her as a rare specimen, there was no way she had fit into the category society deemed as normal. And yet, all she ever wanted was to be normal.

For me, she'll always be the same seventeen she was when we first met. It's rather easy to fall into the same trap, imagine with startling clarity the same giddy feeling I'd felt around her. She had been safe for me.

'La vie est belle!' She'd exclaimed to me once after she'd returned from her French class and I'd returned from my Spanish one. 'Elina est belle!' She'd been quick to follow with.

Even though I hardly knew a lick of French I could tell the pronunciation was all wrong. Belle was pronounced as 'belly' and the 't' was heavily emphasised. But I never rose to correct her.

And in my memories when she speaks it again I can hardly correct her then.

It became something then, I don't know what but it was something. Kyra would say it to me, over and over again, poring over her French textbooks and dictionaries. Her desk would always be cluttered with notebooks and stationery and whatnot and she'd always rake her hands through her hair, frustrated over the next tense.

'La vie est belle,' She'd repeat, practising her French on me, uncaring of my silence. The first part was always the same. 'Life is beautiful' never changed, not for her.

'Elina est belle' eventually became 'Mon amour est belle'. I never understood the difference but the change was welcome all the same.

Sometimes the rage creeps back up and fills me with wrath at sudden moments. And the Kyra in my memories is still just as beautiful as she's hateful.

And welcome or not, Kyra will always remain my biggest what-if. And I loathe it.

I abhor it.

The school year I'm talking about is vivid in my memory, enough for me to describe it as though it'd just occurred. People are contradictory and sometimes can't help being hypocrites. They talk of freedom, equality and acceptance yet scuttle off like frightened animals the moment anything hits too close to home. Some hide their ease behind jests and others can't empathise with those suffering and make jokes at their expense.

It's easy to believe all this when I find myself being mocked, again, by Dhruv, the colossal class idiot. Sighing, I let my head flop down onto my desk ready to doze off and say goodbye to the world. My classmates blabber around me, excited to make new friends and listen to themselves talk.

'Quiet class!' Mrs Chatterjee calls, as she bangs the duster onto the blackboard. 'We have a new student transferring in so let us listen to her introduce herself and do our best to welcome her.'

My eyelids feel heavy and a yawn threatens to burst out as I zone out. I'm ready to sleep through the first period then and wake up for the second.

'Any volunteers to show her around,' Mrs Chatterjee asks, indicating the end of the newbie's intro. I keep my hands down. If I wanted to get to know this one it'll be at my own pace.

'Ah, thank you, Noor,' Mrs Chatterjee speaks again and my urge to groan intensifies. Of course, that goody-two-shoes had volunteered and now my best friend would drag me along for the free ride. It's easy to picture the big doe-eyed look she gives any teacher, looking like a wide-eyed little brown owl, albeit an adorable one.

There's no choice but to at least know what the newbie looks like so I lift my gaze.

The new girl's pale, well pale enough for an Indian to stand out although not beating my Ammi-jan on the fairness scale. She's looking around the room curiously, occasionally brushing light brown strands away from a thin face. I'm leaning forward before I even realise it. Her skin'll probably turn an interesting shade under the sun; whether it'll turn a delicious brown or a rustic golden like Midas had birthed her is what interested me.

I'm still looking when pale green eyes meet mine and hold them until my gaze is sheepishly lowered. It isn't my desire to acquaintance myself with someone new. There's an intense stare on me that prompts me to look up. The epitome of purity smiles at me, pink lips pulling back to bare her teeth in a snarl. And it's easy to fall hook, line and sinker.

'Screw you,' I mutter to Noor, who's sitting next to me, startled. 'What's her name?'

'It's Kyra,' She replies, peering at me in concern with big-brown eyes and soft features. It's impossible to stay mad with this one. 'You good Elina?' I flop down on my desk yet again and groan in response, hoping it translates to "I just had the weirdest feeling that has something to do with the new girl and I don't know what to do with it so thanks for forcing me to spend time with her."

I think not.

All I know is that the moment the bell rings, I'm out of here, best friends or not.

Fuck my life.

Fuck my decisions.

There's music blaring around me and people dancing way too close to be comfortable. It's hard to recognise Noor's usual house in the blinding disco lights, amongst the various tables lined with some drinks. I don't touch them; I may be an 11th grader but I'm not reckless enough to try something that could be potentially spiked. Avoiding people to the best of my abilities I look for the social butterfly that's my best friend.

The air feels heavy and humid and I brush down my jeans and t-shirt uncomfortably. Kids my age are stumbling around smelling of booze and puke while some make out on the couches. It's a wonder that Noor's parents are so lenient, but they trust her to make good decisions. Scanning the rooms makes it clear that Noor isn't there so the terrace is the next best option.

Heading towards the stairs brings me closer to the nearby shouting. I turn the corner and immediately want to back-up. Dhruv's standing there, the same dumb smirk on his face, as he either hits or picks on some poor girl. Knowing his history with me, it could be either. I'm about to skedaddle and look for someone to stop the incoming conflict when he surges forward and swipes at the girl.

Oh, shi-

'Hey!' I shout, hurrying towards the group. 'What do you think you—'

I'm cut off by the neat punch that lands on Dhruv's nose, sending him stumbling backwards. I stop, gaping momentarily as Kyra curses and violently shakes her rapidly darkening hand. Everyone in our immediate vicinity is shocked speechless as they all stare at the new girl. Gathering my wits, I grab her hand and haul both of us up the stairs. The only one who'd care to drag the fight on would be Dhruv, the rest couldn't care less.

'What do you think you're doing?' Kyra demands, allowing herself to be pulled along. 'I was gonna teach him a lesson-'

'The only thing you were gonna teach anyone is how to get suspended within three days,' I cut her off as we climb up the winding sets of stairs. 'Picking a fight within your first week as a newbie has gotta be a new record.'

The answer seems to anger her further and once my grip is loosened she yanks it away and stares furiously at me. Her pale cheeks have reddened along with the tip of her button nose as her shoulders heave up-and-down. Cute.

'So are you saying that I should've let him off?' She demands yet again and I deadpan.

'No, it's good that you let him have it,' I respond, briefly silencing her. 'But it wasn't smart to try and continue the fight because it would've blown up and

Noor's house could've been trashed and I would've gotten involved in a fight. I hate getting involved in fights.' Kyra shifts away from, looking sufficiently chastised.

'Whatever,' she mumbles. 'I just wanted to be a normal transfer student anyway.' She begins to walk away.

'By the way,' I call after her. 'Keep your knuckles facing upward and palm down next time. Drive upwards and use your dominant hand to attack while protecting your face with your left one.' She twitches briefly in acknowledgement but continues forward.

'Elina!' A blur squeals and jumps onto me. 'You finally got here.' Noor heaves a sigh of relief. 'I was getting bored to death without you and some drunk idiot nearly fell off my roof!'

I smile wearily, knowing that getting a word in while Noor is ranting is a near-impossible task. Looking around reiterates what a good job she's done redecorating the house.

The terrace resembles a forest with a green carpet turf with potted plants in every nook and cranny. The Christmas tree in the corner stands awkwardly out of place, its shiny ornaments dangling in a meticulous, orderly line.

My lips twitch at the reminder of Noor's fixation with the precision of objects; my feet ache with old pain, reminiscent of when she stomped on them for accidentally spoiling her Rangoli. Best friends; what a pain.

'We're playing Spin the Bottle now,' She speaks in hushed tones. 'Wanna join?' I shrug and let myself be guided to the circle of people sitting on the ground.

Grabbing two chairs, Noor seats the both of us at the edge of the circle. Smiles are shot my way and my mechanical one widens in response. Being around so many people is stifling. The sky's dark enough for it to be well-nearing midnight and my head slumps onto Noor's shoulders.

"—Lina! Elina!" The shout startles me — I rock backwards in my chair, falling until I'm steadied by arms pushing the chair upright behind me.

"The hell!" I spit harshly, head whipping towards Noor who stares back at me with apologetic eyes. Kyra's face appears in my peripheral vision as she squeezes my shoulder reassuringly and pacifies Noor.

"We called her but she wasn't responding! I didn't mean to scare her. Sorry about that Elina, but I was calling you for a long time," Noor defends.

'Right, right okay,' I murmur. 'So it's my turn?' A few "yes"es resound and Noor grins at me mischievously. 'Who do I have to kiss?'

A pale arm shoots up and Kyra smirks at me.

Oh hell no.

'Am I allowed to withdraw?' I press, avoiding all eyes. 'Do two girls have to kiss?' A few uncomfortable giggles start and quickly die out. Some of the kids look maliciously gleeful as they pull their phones out.

Great.

I chance a look at Kyra and blanch. Her expression is almost entirely shuttered off as she looks at me blankly. Her eyes look cold and frigid as her mouth is upturned in disgust. Guess even she isn't fond of this kiss.

I begin to protest again, unwilling to give my first kiss away so easily to someone I didn't even like, a girl at that. Noor sticks up for me, convincing the rest to put down their phones.

'Come on,' A girl with blue highlights insists. 'It's just a kiss, we all participated.' Okay so the blue-hair was sick but the girl was a little shit.

'If you wanna experience this so badly then do it yourself,' I respond, irritation coating my words. 'I don't swing like that so it's not cool.' Blue-haired girl flushes and falls silent.

'Did you say "Not cool"?' Kyra asks mockingly. 'What part of this could be cool if I have to kiss such a big baby? This isn't normal for me either. Stop making such a huge deal out of it and let's get it over with already. Then at least you can upgrade from the first-base virgin.' I blink, surprise and embarrassment flooding through me.

Noor rises, curses at the tip of her as my head begins to pound heavily.

'It's 'just a dare' and why does it even matter?' Kyra continues. 'Who cares about first kisses and all that crap? If everyone else can take part in the game why's Elina the only god damn wimp here?'

My fists clench tight as I imagine slamming my fist into her jaw and hearing it shatter as hot tears gather in my eyes. As a chorus of 'Preach!' follows the girl's pretty little monologue, I begin rising from my chair. I know I shouldn't fight her here and damage Noor's house but the temptation is simply too hard to resist.

I'm up now with nails digging into my palms sharp enough to draw blood and I don't want to disappoint my family and my friends again.

A gentle shove makes sit back down and I catch Noor's eye as she slowly shakes her head. Her concern snaps me out of my rage. Although the hype has died down, like pretty vultures all dolled up, the little clique hones in on me expectantly and I ignore them.

'What the hell Kyra!' Noor screams, a red flush developing on her face as she stomps closer to the other girl. The girls around Kyra flinch and back away — despite her sweet disposition and rare temper, Noor does practice boxing, so the apprehension is justified. Kyra looks unruffled even as she is met head-on.

'Why the hell would you say all that?' Noor shrieks, getting right in Kyra's face. 'What makes you think you have the right to insult Elina you little bi—'

'Stop,' I direct. 'This has become far too dramatic. We'll just kiss and get it over with.'

Noor begins to protest yet again until I silence her calmly. Walking closer allows my eyes to fall onto Kyra's lips; thick, plump ones a soft pink. My face is an inch away from her grin, so my eyes close automatically. The pressure is soft and the breath in my mouth is sweet;

the kiss goes on for a few seconds and I don't realize I've blacked out until I feel something slimy nudge my tongue piercing and immediately recoil.

The two of us separate and I thank every existing God, deity, supernatural being, whatever, that I don't even blush.

Kyra's green eyes peer into my own steadily and she slowly licks her lips. My eyebrows nearly vanish into my hairline at the tacky move. My incredulity must be easily visible because Kyra chuckles and leans forward.

'Not bad for a first-timer,' She whispers. 'But I'm up for more practice if you are.' I gape and then hastily back away. The rest of the game has a strange sort of tension and I'm left wondering about the scent of the sweet lilac flowers and taste of cherry and wanting to go home. Obviously, things go wrong.

'Where's Kyra?' Noor asks me, looking around curiously for the newbie as her guests begin to trickle out at around 2 am. I shrug, somewhat spooked by her pissed off expression. A guy walking past us turns and asks us if we're talking about the newbie. At our nods, he brightens.

'Oh, Dhruv called her for something so she went after him. They're at the entrance' What...

'What the fuck!' I shriek, breaking into a sprint and ignoring Noor's surprised shouts. 'Does she plan on getting expelled or beat up because of her bravado?' I race at full speed, shoving past the exiting people as they swear at me. Soon enough Noor falls into step behind me as we race out of her house

The shouting draws our attention and puts me ahead of Noor quite easily, my leg muscles recalling all my exercises and training. The scene right at the entrance of the house draws a scream from my throat. It's easy enough to see Kyra being herded by at least four people one of them being Dhruv. Before anything else happens she pulls her fist backwards and then sends it flying right into the nearest guys' jaw. Her left hand is defensively curled around her face and her right immediately snaps back to attention. The guy tumbles into Dhruv and another girl who steadies him.

'Why you!' He begins to scream, advancing on the shorter girl when my open palm strikes his nose, sending him off-kilter. I pivot, elbow tucked inwards and then punch him with a clenched fist, landing my hit on the same spot as he sits down heavily. Behind me Noor moves swiftly, using her speed and small stature to her advantage and takes down the girl and another guy. Kyra is repeatedly dodging hits from the fourth guy but it's easy to see her exhaustion become apparent. I move towards only to stumble backwards as Dhruv surges up and punches my chest.

I gasp, winded as he head-butts me. Nearly stumbling, my hand grabs his arm and remembering an old move, my leg sweeps out his as he collapses onto the ground. My other hand grips his shoulder and a little pushing will break the bone. My head pounds and begins to feel hot and I can do it—

Then there are arms pulling me back as the students scream at me to calm down. I collapse without struggle and stare incomprehensibly ahead. I nearly lost my control today. Dammit.

I can't sleep much that night, too confused by something or the other to doze for more than an hour at a time. Propping open my laptop I binge-watch the entire season of Brooklyn Nine-Nine till I'm hungry enough to hunt for a midnight snack. Rummaging through the fridge, silently to avoid waking my parents, I grab the cherry cake left and gulp it down. A few minutes after finishing I feel guilty and poke my stomach idly while smacking my lips. The taste of cherry is pretty prominent.

My mind flashes back to the kiss and the sweet sensations bursting in my mouth. What did Kyra even mean by that? Did she like me? Was it a joke? The kiss replays in my head again, leaving me slightly giddy and mostly confused.

I didn't dislike the kiss.

I bury my head into my blanket and roll around on my bed, screaming as Avengers: Age of Ultron plays in the background.

A smug expression and pouty lips flash across my sight.

Oh. I didn't just not dislike the kiss. I liked it.

Well, shit.

"Language!" Captain America chides.

Now, I don't take extreme actions in most circumstances, but my philosophy is to either go big or go home — I identify a flaw in this ideology as I find myself trying to tiptoe into my house after school and the looming days of suspension— maybe my extremism doesn't always result in the best decisions. As my luck would have it, of course, my elbow strikes the door frame right before I step inside the apartment.

Brilliant.

I curse loudly, clutching the wounded joint as a sharp pain shoots through my arm, momentarily numbing it and then exposing me to a whole new world of pain.

'I heard that!' My father calls disapprovingly, leaning against the doorframe, unimpressed with a raised eyebrow.

I grin sheepishly. The scent of oil and roasted spices frying meat wafts in the air — 'Chicken biryani?' I ask, a desperate attempt to cut through the tension.

His glare unrelenting, he turns around and heads back toward the kitchen, and I follow, stumbling as I struggle to keep up with his silent beckoning.

We enter the dining room to the sight of my mother and sister sitting at the brick-red dining table, scarfing down what remained on their plates as their voices rose above each other with talk of the latest developments in the news, the opening of a club next to our home and the new Ranbir Kapoor movie all in one breath.

Ma's still wearing her office suit, a mixture of white and black. The combination is enriching against her coffee brown skin that I take after along with her slender and curved features. My curly locks, however, don't resemble her straight locks. She, like my sister, is wearing thick black glasses.

Diya takes after Pa. They're both a tanned golden with short, crisp hair only hers in a pixie cut. Her and Pa's features are thin and long. The only thing we have in common is our russet brown eyes. The one thing I have in common with her is loose and comfortable clothing.

The sun is still high in the air outside and any remaining sunlight is blocked by the light gives the garnet red walls and the scarlet tables, chairs and sofas a sombre look. I attempt to slip to the bathroom in vain, my bag crashing onto the floor and catching me mid escape. Sighing, I take my books out and place them into the black bookshelf standing proudly as the only different colour in the room.

"I'm back," I say, waving wildly in response to the chorused greeting. Silence falls.

'Elina,' Ma begins. 'We got a mail from the school saying that you've been suspended for a week after you fought with one of your classmates. Arjun and I have to go to your school tomorrow to talk to the Principal.'

Pa grunts in acknowledgement. Diya's shocked gaze turns from me to Ma and back again.

'I know,' I mumble, slinking back into my seat and stifling the uneasy churning in my gut. 'Dhruv started the fight, I'm just sorry that I lost control of myself and dragged it on'.

Pa comes out of the kitchen and sets down two plates of steaming, hot biryani and removes the apron around Ma's waist.

'Tell us what happened.'

He settles down at the table and we begin our daily dinner.

'And don't try to make excuses.' His hazel eyes wink at mine and I sigh, assured that I will get a chance to explain myself.

And so I tell them about the entire incident, omitting the kiss out entirely. Diya looks faintly amused upon my admission of how my time as a boxer had perfected my punches. Pa's watching me closely, an unreadable look in his eyes while I proudly describe how Kyra followed the tips I gave her and nearly knocked a guy down. When I'm done, Pa promises to handle the situation and goes to talk to Mrs Chatterjee. Diya claps me on the shoulder and congratulates me on a job well done.

Soon enough, it's just Ma and me, sitting in silence. She gets up from the table and quietly asks me to follow her. We enter the privacy of her room, the master bedroom, and sit on the champagne coloured bed sheets.

'Elina,' Ma asks, drawing my gaze away from the Egyptian blue walls and the simple white furniture. 'Do you think you're being unable to control your temper? Do you want us to seek outside help?' I tense, head throbbing because I don't want to go to therapy and 'talk about my feelings'. I stare stubbornly at the pale blue marble floor, causing her to sigh.

Ma reaches over and grasps my hand.

'Elina, the incident at the boxing club bothers you more than anyone else,' She tries, oblivious to how I drift away from the conversation. 'You can't have a repeat like that and we ought to help you with-'

'It's okay Ma,' I repeat. 'It's fine so you can go and sleep now.' Ma searches my face, and apparently what she finds is enough for her to back off. For now.

The moment I enter my room I collapse on my bed, burrowing my head in the blankets. The incident at the boxing club had gotten me kicked out and earned me a one-way ticket to therapy. That day is the most revolting day to exist for me. A sigh escapes and my eyes close at around 5 pm.

I return to school next week and attempt to avoid Kyra — considering that we hardly know each other, it isn't particularly difficult either.

Skulking around the cafeteria during the first couple of minutes grants me a plate full of white pasta that I lick clean in seconds. Noor suffers the unfortunate fate of listening to my complaints about the lack of seasoning and whatnot in the cafeteria food while being dragged around.

And because I need to suffer as well, we stop every 5 minutes as yet a new person approaches Noor eagerly. When we finally grab a table, it's to some guy a grade below us regaling her with a tale about seeing a real tiger in the wild, when Kyra slumps down onto our desk.

'Did you want something?' I ask, irate expression manifesting automatically at the sight of her.

Kyra raises her brow; 'I can't sit with my friends now?' The mockery in those words is obvious and something snaps.

'Considering that we met a total of maybe two times and you managed to piss me off in that time-span, then no, we're not friends who sit together,' I snap. 'And don't bother including Noor in this because she's almost as pissed off.'

Kyra looks impressively shocked. The change in my usual lacklustre attitude could give anyone whiplash. She blinks a couple of times in rapid succession before my unimpressed face as her fidgeting nervously. Glancing between the two of us, she takes a deep breath and sits up straighter, meeting my gaze head-on.

'I'm sorry,' Kyra says directly. 'I shouldn't have been so callous and rude to you when you just did what most people would do and responded normally. And it wasn't my place to talk about your love-life or anything like that. And also, thank you for helping me out when it was my fault.'

The strange wording confuses me but the apology is honest enough for me to accept it. Upon my acceptance, Noor does the same and we eat in silence. After we're done, Kyra asks to see our timetables and compares our subjects. We have all the subjects except our Second language, performing arts, arts and crafts, and sports together. Kyra will head for French while Noor and I'll go with Spanish. Sports are divided according to roll numbers so Kyra and Noor have a greater chance of being grouped.

We begin to walk in our separate directions with Noor leading me forward when Kyra calls after us, asking if there's something else I should know.

'Oh yeah,' I answer back. 'You'll meet this girl called Aditi if she isn't absent. Go over to her and say that I sent you, she'll help you if you're confused anywhere.' And knowing Aditi, she'll be eager to help out.

A couple of determined knocks sound resoundingly on my door and a muffled groan comes out in response. There's the distant tone of someone calling my name but it's too early for this bullshit.

'I didn't cancel your Netflix subscription on purpose Diya!' I call, hoping it'll shut her up enough for me to catch five more minutes. There's silence on the other end but then the knocking begins again.

Cursing, I swing my legs out of the bed and nearly fall. Stumbling forward while wrapped in a blanket, I pull the door open with a very annoyed 'What?' Kyra stands there, trying and failing to hide a smile as she looks me up and down. Suddenly very aware of the tangled mess that is my hair, I trace my fingers through the strands desperately, wincing whenever I pull a knot. Kyra snorts and steps forward, helping me out by smoothing the strands. My hands fall as I stare at her face.

There's a wave of freckles across her nose that I want to count. Her brows are furrowed in concentration and nose scrunched up prettily as she works on brushing my hair next. Her eyes are lowered and covered by thick lashes so it's hard to see them. My eyes fall onto her lips next and my face begins to burn. I snap my gaze upward and dodge out of her reaching hands, grumbling about troublesome behaviour. My ears feel hot and I know that if anyone looks at them, they'll be bright red.

'Dude,' Kyra says, gaze trained firmly on my wrinkled uniform. 'Did you sleep in the school uniform last night or something?' The white skirt is crumpled at the edges and the beige t-shirt has undone buttons. Grunting, my hands sweep over the outfit absentmindedly.

'Or something,' I respond. 'I woke up at 6, got ready and then dozed off again.' Grabbing my bag, I begin stuffing my books inside. My skirt rides up high and a shiver runs down my spine at the sudden burst of chilled air from the AC. Nearby, Kyra walks further in and curiously observes the punching bag attached right in the middle of the room. Moving forward, she lets out appreciative noises as she surveys the teal green walls with little wooden trinkets hanging from them and the black and white decor. As I zip my bag up, she moves onto wearing a wooden Inca mask she plucked off the wall and plops onto my grey armchair.

'Did you paint these?' Kyra asks, and I stop momentarily to see her point toward the painting on one of my walls. A large black butterfly is painted right above the headboard of my bed, its' wings stretching far and wide. I shake my head.

'Nah, my dad did. He's a professional painter.' Kyra lets out a gasp, awe adorning her face. Propping my leg up, I begin to tie my black shoelaces.

'Hey, Elina you're...' Kyra trails off momentarily, voice hoarse. 'Your underwear is kinda visible.' Humming, I prop up my other leg and begin the process again.

'That's fine,' I say. 'Just don't look.' Finally, my shoes are secure so I race for my bag.

'I wasn't...' Kyra begins. 'No one would- You know what, never-mind.' Letting out a subdued moan she trails after me as we walk to the dining area, say goodbye to my parents and then race out the door. The ride to school is filled with a lot of chattering about something or the other. Seeing our first period is Second Language I grab onto Noor and bade goodbye to Kyra.

'What's with you?' Noor asks as I tug her around, grinning like an idiot. 'Since when were you and Kyra on the same bus?' I shrug, replying that it just has to do with us coincidentally living in the same society and all. Noor eyes me suspiciously but has to drop the subject as Ms Esperanza begins the lesson.

After class, I'm packing up when a low call of my name catches my attention.

"Hey hon," Aditi croons, voice husky and purposely sultry, winking bright brown orbs at me. She reaches around and snakes her arms around my waist, placing her dark head just a couple of inches below my chin. Her toned body presses flush against mine and feline features smirk at me.

"Yo darlin'," I snark back, adopting the accent that sounds the poshest to me and butchering an Australian accent.

I ease myself out of her grip and smile happily.

Her figure is strict and skin like ice-cold coffee but her easy smile reveals more than her greetings ever would. We catch up quickly with each other after nearly a week of not bumping into the other and question any significant event that occurred.

'So, I heard from the newbie that you got suspended for trying to be her valiant knight in shining armour?' She quips without any real heat. 'And how come this random chick knew about everything that happened to you from the beginning of this term?' Her arms squeeze my middle and she sulks.

Laughing I remove her arm and pinch her nose; 'Save the drama for your theatre club bro. That's the only time you'll get any attention.' Before she can snark with a witty comeback, Noor descends on her, wailing about being ignored and whatnot. While Aditi's busy placating my best friend, I hear a low cough from my right side. Swivelling around, I see Kyra dump all her French textbooks onto a random desk and slump onto groaning.

'French sucks and I'm never going to use Plus-que-parfait in my life!' She exclaims woefully until her eyes meet mine. Kyra clambers up and rants about a new phrase that she learnt. 'La vie est belle, Elina!'

I grimace, certain from my limited knowledge of French that the pronunciation is wrong. Thankfully, Aditi decides to be my saving grace;

'It's "La vie est belle" without the "t" emphasised and the "belle" not pronounced as "belly" you dumbass,' She scoffs. Kyra, in turn, glares at her without any restraint. I frown. Aren't these two on good terms? I thought Aditi would rather like Kyra and vice-versa. From Noor's startled expression she didn't expect this visible hostility either.

Bidding the rest a hurried goodbye, I trudge off to the dance room, feeling a strange kind of excitement I haven't felt since boxing. Knocking on the oakwood door earns me a brief "Come in!" as I slip inside the room. The pirouetting dancers all greet me happily and shift themselves to open up a spot for me. I drop my shirt on the floor and stand in my tank-top and shorts, facing the giant mirror extending from both the ends of the third wall. I begin to pirouette in succession, showcasing a wilder side to the softness of ballet. My legs make soft 'thumps' on the wooden floor, the material creaking lightly under my feet.

By the end of my last spin, the two beige walls have all blurred with the grey ceiling and so I stop. The memory of the boxing incident is still clear in my mind. The last wall opens out to a playground and usually, a curtain separates the two areas, something absent today so I feel the cool air on features and in isolation, surrounded by dance and acceptance, I am at peace.

Exiting the dance room after the last bell has rung, I spot Kyra standing there. I grab Kyra's hand and tug her along. She lets out a startled squawk but lets me pull her down the dimly-lit hallway. We walk toward the school gates leisurely, talking aimlessly.

'How're you finding school so far?' I ask, looking at our joined hands and beaming. 'Any classes you enjoy? Teachers you hate or guys you like?' Kyra shrugs, yawning dispassionately.

'St. Alexius is a better school than my last one,' She admits. 'I didn't have a lot of friends and my ex was a douche. Plus, this place is closer to my parents so I get to live with them now.' Kyra squeezes my hand absent-mindedly, unaware of my shock.

'You didn't live your parents in the past?' I say, aiming at nonchalance. 'Pretty busy with work or something?' Kyra shrugs again. Unlike previously, her eyes are hardened little orbs. Her hand twitches in mind momentarily, drawing my full attention to it.

'Sort of,' She responds tersely. 'I lived with my Nani and cousins in Lucknow while Mom and Dad are based here in Delhi. A few things came up so they decided to switch my school and my parents agreed to take me in.' My eyes flicker over to her face, observing as she glares into the distance. The look on her face can easily be called thunderous with tightly-knit eyebrows, a fierce scowl and darkened eyes.

'What came up?' I press. The topic displeases her but why? I wait till we've found our bus and then clamber atop it. Immediately I chuck my shirt at the last seat and call dibs, grabbing the window seat. Kyra groans and reluctantly lets me get in first, sitting down next to me.

'The ex I mentioned,' Kyra sighs. 'He started some dumb rumours about me being depressed and my parents being absentees. And all the kids just swallowed it up and became little assholes. The teachers were pretentious as well, talking about feelings or some shit. So I left.' I nod. It's an instinctual feeling that she doesn't want to tell me what the rumours were but leave it be.

'I always try to fit into "normal" groups and mingle with everyone,' She continues. 'My parents are the most boring beings ever. All they care about is how society views them and how they need to be the most normal parents out there. So I get good grades, play a few sports, have good friends, don't drink, don't have a boyfriend, don't do anything considered unnatural.' She attempts a smile and the sympathy I feel for her comes pouring out.

Everyone wants to be accepted by society and be normal. And we often hide our true selves because of it.

We sit in silence, looking at the kids clambering on until the bell rings one last time. There's a sound like a gunshot and a bus begins to rumble and wakes up. Looking through the window, a dull blue sky greets me. The bus is rolling past a trash heap with dogs sniffing eagerly and cows munching on something out of the plastic bags. The sun's well over our heads, setting steadily into the west. My eyes shift from huts to mud-houses until a low voice breaks me out of my trance.

'Do you mind answering something personal?' Kyra asks, meeting my eyes on the window screen. I hum, staring at her in return.

'During our fight the other day when you'd pinned down Dhruv, you looked ready to break his arm in two.' I nod, already disliking the direction of the conversation. 'You almost did until those students pulled you off and you went limp and blank, as though you weren't even there. And at Noor's party when we first met, you said you hated conflict even though you're a great fighter. What's up with that?'

I stay silent, dropping her gaze and staring outside the window again. My fists clench and unclench as I open my palms to reveal a few scars. A sudden shuffling next to me causes me to flinch away. Kyra raises an eyebrow and drags my hands onto her lap. Flipping them so my palms uneasily grip her thighs, she runs her fingers lightly over my knuckles. The gentle brushes cause me to shiver as I fight the urge to yank my hands away.

'I used to box,' I blurt out, falling silent when Kyra turns to look at me. 'I joined when I was around ten at this academy near the society and went four times every week. I convinced Noor to accompany me and she joined when we were thirteen. I practised competitively and was pretty sure two years ago that I wanted to be a boxer.' A sigh escapes me.

Kyra rubs my knuckles comfortingly and stares at the light abrasions disfiguring the otherwise smooth, brown skin. The marks are recent, courtesy of the punching bag in my room nearly getting annihilated. My hands twitch as I make the mental reminder to wear the protective gear next time.

'Last year though, I was pretty... unhappy,' I continue, closing my eyes and tilting my head back onto the rough leather seat. 'Boards put a lot of pressure on me academically, my friends and I were too busy to talk to each other and I was aiming for nationals in boxing. But the mood swings were sudden and violent.' It's easy to form an image of that day, as I screamed and whaled on the punching bags after a violent argument with Ammi-jan.

And then I'd had a match against one of the juniors. They were some usual trash-talk going on in the sidelines which had made me even more pissed off. We'd circled each other, my fists clenched at the side and my nostrils flared. The bell had rung and I'd leapt. The girls' hands had come up to protect her face leaving her stomach and ribs exposed. Amateur move. I aimed an uppercut there, driving my clenched fist in hard into the soft skin. She gasped, arms automatically lowering to clench her stomach and my next fist landed clean on her temple, knocking her down.

I screamed, drawing my foot back and kicking her in the ribs. Blood pounded in my veins. 'Not good enough' Ammi-jan whispered, the ice-cold words piercing my skin. My foot drew back, ready to strike when two arms clasped around me and pulled us apart. Thrashing, I elbowed backwards, earning a winded gasp and then punched the face next to mine.

'Elina!' The coach had roared and I finally stopped, turning to look at the face beside mine. Noor stood there, shaking and holding her bloodied nose. She grimaced and then tilted her head down, preventing the blood from choking her. My breaths came in stuttering gasps as I surveyed the damage I'd caused. My opponent was on the floor, holding in tears as Coach Singh carefully propped her up, attempting to wipe her face clean of blood and gently pressing against her ribs.

My eyes wandered back to Noor and the red swelling of her face. And it's Noor that I'd hurt. The beast in me calmed down and tears filled my eyes.

Kyra looks at me wide-eyed, as though unable to picture me hurting Noor or anyone else so viciously. But it's not hard to connect me to the beast who nearly broke Dhruv's arm.

'I went for therapy then, for almost a year,' I continue, voice thick and eyes, that I don't recall opening, blurry with tears. 'I've gotten a better hold over my temper and quit boxing. But sometimes, like with Dhruv, it's easy to slip up.' The sight of Noor's face being splattered with blood as my punch nearly broke her nose, the way the girl had clutched her bruised ribs and then puked over the side of the rink is still clear. I revisit the memory over and over again to remind myself of the consequences of my anger.

My eyes close again and I let out a shuddering breath. Kyra's roving my features, her pale gaze heavy. Drawing in a breath feels painful, my ribs seem to shudder and my knuckles throb with phantom pain. She shuffles a bit, hesitant and unsure but stays quiet. She says nothing, even as the silent tears roll down my cheeks and my mouth parts to let out a small whimper. My shirt is removed from around my waist and tossed haphazardly onto my face, shielding my grief from the moving bus. I clutch the sleeves desperately, wetting the material with my tears.

As I drift off, exhausted from both the day and the revisitation, Kyra gently holds my hands, drawing circles over the split skin of my knuckles. And despite the grief, something warm settles in my chest

Pa comes to pick me up from the bus station, standing by my side as I bid goodbye to Kyra wistfully and tells me to follow him to his 'art studio'. Obliging, we walk into our home out onto the main balcony overlooking our society's park and the Aravalli hills, that we call 'the art studio'. It's a large balcony, around five feet deep and ten feet wide, with partially recessed glass railings. All sorts of canvases lie either scattered around or propped on top of easels. A box full of acrylic paints, paintbrushes, a palette, a bottle of satin varnish and charcoal pencils is carefully situated at the centre of a wooden table. A small stool has a rag and Pa's apron tossed on it. At the very corner of the balcony, a jug of water with soap for clean up rests.

'I seem to have lost my palette knife and masking tape,' Pa grumbles, tying his shoulder-length hair into a small ponytail at the base. 'Diya'll bring the scraping tools so sit down somewhere and tell me about your latest art project.' I plop down next to the jar and describe the theme of tragic love for the art class. Pa hums, mumbling a few names here and there. 'Romeo and Juliet' is shot down by both of us quickly, opting to go for a more mystical feel.

'How about we go for Greek tragedies for this one,' Pa suggests, chewing his lip thoughtfully. 'Orpheus and Eurydice?' I shake my head, a tragedy indeed but it doesn't feel right.

'How about one related to Apollo?' I offer, the silvers of a story coming to mind about the tragic love of the Greek God of Music. Pa clicks his fingers.

'You mean his first love the one who turned into a tree,' He proclaims. 'Apollo and Daphne.' I pause, thinking of the sad tale of unrequited love and the role Eros had played in it. A good tale, but something is missing. I shake my head yet again.

'Not that one,' I reject. 'I mean the tale of Apollo and Hyacinthus.' Without waiting for an answer I move forward, brush at bay and canvas gripped firmly in one hand. Pa lets out a contemplative grunt and then moves next to me. We work in silence, my hand steady and my mind wandering away to pale green eyes and a sharp smile.

The sun has very nearly set, with the sky a darkened horizon when we step back to admire my piece. It's the infamous farewell when Apollo holds his dying lover in his embrace. The mortal's head is bloodied and wounded by the jealousy of Zephyrus, the otherwise gentle Greek God of the West Wind. Hyacinthus, the man both Apollo and Zephyrus dies to give rise to the first Hyacinth, a beautiful flower created from Apollo's grief. The three figures are present, with Apollo holding the dying mortal and Zephyrus watching as the west wind, but there is a certain femininity in their delicate features and gentle bodies.

I set the brush down and look critically at the work. The painting is that of an amateur, leaving much to be desired but it's mine. Apollos' eyes are dark brown whilst the dying Hyacinthus has pale green eyes. Zephyrus' eyes are endless voids.

'It's a good painting,' Pa says, peering closely at the three figures. 'Did you have someone specific in mind while drawing this?' My mind immediately wanders to Hyacinthus' eyes and shrug. Pa hums, eyes roaming all over the painting.

'The way you painted the three by giving them a certain delicacy and femininity was also good,' He comments shrewdly and then moves onto critiquing my work. If I slip up a few times and name my inspiration, Pa stays silent and encouraging. That never changes even when I'm describing the homosexual romance and its' significance. And it seems like it's abundantly clear what I feel now.

That night, sitting in the dim, hypnotic light of the orange fluorescent light tube, I open my laptop and open Incognito. Well-aware that Google does, in fact, not hold the solutions to the problems in my love life (or lack thereof), it's not exactly like I have a better alternative. Besides, when in doubt, ask Google Baba.

'I think I might be a girl who likes other girls but I'm not sure and I still like guys,' I type in. 'Could I be bi?' I hit search and immediately roll my eyes at the numerous quizzes that pop up. Useless, every single one of them. Scrolling past articles, a link catches my attention and I click it. Immediately I'm taken to a website with the word 'Bi-curious plastered on it.'

"Bi-curiosity," I read aloud. "This refers to people interested in a relationship with the same sex, or at least the experience of it, without labelling themselves as bisexual or homosexual."

Is that even a thing? Dating a girl isn't a new concept, I just never thought it would apply to me of all people.

Sighing, I set the laptop aside and flop onto the bed. Unable to let the glaring imprint of the screen fade away, my eyes are wide and staring. Deciding to sleep on it, my next thought is to not be so quick to label my sexuality and give myself both the time and experience. But, I remind myself, do make a move on Kyra soon.

Anger-II

There's a certain amount of delicacy some parents treat their children with, keeping in mind that these young impressionable minds shouldn't be exposed to the complicated matters of sexuality, sexism, homophobia, racism and the likes. And then these same parents expose these young impressionable minds to religion, ethnicities, societal norms, gender roles and the likes. My aunt, or Ammi-jan as both Diya and I called her, was Ma's sister and almost another mother. She'd been around for as long as I can remember, cheering us on from the sidelines. Ammi-jan is married, but her son died a long time ago in a car accident. I can barely remember him, Rohit he was called, being a mere four at the time.

After the accident, Ammi-jan didn't visit for nearly two months after which she just stopped by one day. She never let Diya or me out of her sights again. I'd been estranged with her for nearly a year, having cut off all possible means of communication after our frequent fights. Hence my surprise when I see her just sitting at the dining table nearly two months after the start of the eleventh grade.

'Hello Elina,' She greets, inclining her light head in my direction. Ammi-jan's wearing a peach coloured kurta with her dupatta tied securely across her hips. Unlike Ma, Ammi-jan has pale blue eyes that take after my Nanu and a light tan to go along with it. She pushes large, red sunglasses into her black hair and beckons me forward. 'Come and give your poor Ammi-jan a kiss.'

I look questioningly at Diya who nods and move forward. Hesitantly, I wrap my hands around her shoulders, grip breakable at the slightest moment of unease. Clucking her tongue, she pulls me forward and squeezes me tightly.

'You've gotten so big since the last time I saw you,' She murmurs and I roll my eyes. I grew a whopping inch from 5'7 to 5'8. Regardless, I have to bend forward to let her hug me.

'Geeta,' Ma mutters, looking at my unamused expression and then at the clock. 'Elina will be late for the bus.' The clock read 6:50 in bold, red digits. The bus doesn't arrive for another fifteen minutes.

'Of course Parul,' Ammi-jan agrees, releasing me easily even as her hands all but dig into my flesh. 'I'll continue my discussion with Elina sometimes later today. We have all week.' She flashes me a smile, lips nude and mocking. I stare back, unnerved. Pa's sudden throat-clearing makes me rush out of the door. But not too fast, I remind myself, or it'll be like I'm running.

Bolting out, a figure almost immediately crashes into me. Reflexively my arm reaches out and steadies the person. Kyra blinks at me, stupefied at my sudden entrance and then lightly taps my arm. Reluctantly, I withdraw my arms and let her step back. Seeing her grin is enough for my present mood to liven up.

'Where's the fire?' She asks, falling into step with me. 'How come you're out early?' I grimace, not wanting to lie to her but unwilling to tell her everything, at least for now.

'My aunt, Ammi-jan I call her, visited and I kinda woke up early for that,' I offer, choosing not to tell her that I wake up fine on my own. Her coming into my room is just a bonus. Kyra silently assesses me, but her concern must win over her curiosity because she offers me a wan smile.

We walk the rest of the way in silence as I think about the implications of why Ammi-jan could be back. Unpleasant thoughts swarm my mind, nearly all of them including the girl walking next to me. Until she gives my hand a comforting squeeze I don't know when I've tightly gripped hers'. She looks at me, previous weariness gone and face full of frenzy. I watch her eyes crinkle up like they always do when she smiles and I want to kiss her.

'So, you're back in the world of the living?' Kyra mocks without any real heat. 'Thanks for bestowing us mortals with your valuable time.' I wrinkle my nose in disgust and then opt to shove her lightly. She shoves me back and a light catfight begins. Okay, I think, there's no going back. Cool, cool, cool, cool.

Diya walks into my room at 5 pm and gets smacked in the face by my hoodie. Spluttering, she tosses the shirt down and glares at me. Ammi-jan walks in and all my good cheer disappears.

'I have to complete my semester courses by tomorrow,' Diya says by way of apology. 'So, Ammi-jan will be driving you to Ambience.' The beguiling smile on her face intensifies my urge to fight-or-flight. My hands twitch with the usual phantom pain and I desperately eye Diya who offers me a vague, sheepish look but offers no help otherwise. Traitor.

Sighing, I follow Ammi-jan to her car. The silver Maruti is small but large enough for me to sit as far away as possible. In close quarters the sweet, subtle and heady fragrance of flowers issuing from Ammi-jan is apparent. It's the same French perfume Ma wears, the one that smells like how whisky tastes; strong and addictive. Looking at this frivolously dressed woman in a garden hat, it's hard to imagine my bread-winning Ma as her sister.

'It's called "Premier Jour" by Nina Ricca,' she says, black-tinted sunglasses glancing in my direction. 'Both Parul and I wear it.' Her painted-red lips pull into a wide smile and it's easy to realise that appearances can be deceiving. This woman is as astute as ever.

'It's nice to see you finally wear some decent clothes,' Ammi-jan continues. 'At least, for now, you seem to have grown out of the ratty shirt and bland trousers stage.' She nods toward the white boxy crop top I'm wearing and the accompanying black flared out pants.

'Soon you'll be enough of a woman to wear the jewellery I have for you.'

In the ensuing silence, the air grows too constricting. Ignoring the AC, I roll down the windows and breathe in the air. Ammi-jan frowns at me from the rear-view mirror. It's too much to hope she'll stay silent but I hope so anyway. She stays silent for a good 5 minutes though.

'This friend you're meeting,' She begins again. 'Are they a girl or a guy?' I answer girl and she disappointedly shakes her head. I'm confused until Ammi-jan begins murmuring about having a stable boyfriend at this age. A flash of irritation causes me to briefly close my eyes and slowly count to ten.

'It doesn't have to be a guy,' I mumble. 'Bisexuality exists.' Her face visibly tilts to stare at me long and hard in the mirror. Driving to the side and slowing down the car, she takes off her sunglasses, focusing frosty blue eyes on me. Trying to breathe normally, I wipe my face clean of any lingering expression.

'Teenagers these days,' She begins, cool and calculating. 'Love to use big words they don't understand to show off their "intelligence".' She looks away before the indignation crosses my face. 'Just because an American slang is created doesn't mean it applies to Indians.' Incredulously, I glare at her in the mirror.

'Where the term was coined doesn't matter,' I retort. 'The term is a scientific one which has been and is being studied. You can't just pretend it doesn't exist.' Her hands grip the steering wheel.

'Some people like to experiment too much and they need help from others,' She grits out through clenched teeth. 'And if kids hang out in the wrong company then they get influenced by these ridiculous ideas and don't realise it's just a phase.' My nails scratch up the leather seat.

'They don't need your help,' I snap back. 'Homosexuality, bisexuality and the whole umbrella term exists and is present in other animal kingdoms and is viewed by scientists as normal. It's certainly not a phase and one's company has nothing to do with it.' Her hand slams on the steering wheel, drawing out a long horn.

'It's sick and unnatural and these people need mental help and proper guidance,' Ammi-jan screams, spittle flying from her mouth.

'We don't need to be told how we need to live our lives by others!' I roar back, hand punching her seat and drawing out a yelp. 'They don't need someone so closeted and close-minded like you to poke around in their business and then try to diagnose them. So for fuck's sake, mind your own business and don't try to pretend you understand everything.' My hand's throbbing, a bright red and my vision's blurry with tears. I'm panting loudly, a loud ringing receding in my ears and suddenly slump down like I'm a puppet whose strings have been cut.

'Don't try to pretend you understand this after you took away boxing from me,' I whisper. I bit into my cheek, tasting blood and violently blink back tears. I won't give her the satisfaction. 'And before you say that, no, this isn't about my boxing. It's about how you can't understand anyone different from you and you always try to take that difference away.' I look away, snuffling quietly and rubbing my nose vigorously. Ammi-jan drives, quiet and subdued at my sudden outburst.

The moment we arrive at the mall I jump out of the car and walk off. The memories of her belittling me the moment I decided to take up boxing as a career, the disgust for a sport that wasn't even "feminine" and the constant pestering wash through me again. In her eyes, I was a disgrace and a poor excuse of a girl for pursuing my sport. Everything I did was lacklustre.

'You're lazy,' She'd say to me, looking at me after I collapsed from overworking myself with continuous boxing training and academic pressure. 'You're dispassionate, laidback, sloppy, apathetic, careless, shortsighted, irresponsible, overly-dependent, and unable to work under pressure.' That's all I was, a disappointment. And naively enough, I believed her

She'd lied to me once, said that Pa had decided to withdraw me from the academy. That had led to the most hurtful argument I'd ever had with him. And she had the gall to watch and pretend it was "for my own good" when I screamed at Pa through my tears and hiccups and told him I hated him for taking away my dreams. The stricken look on his face makes me so sad even now. And realising that I'd been deceived by my Ammi-jan, my second mom was the final straw. I'd told her to get out of my sight and then left for the upcoming boxing match against a junior.

I'd returned home that evening, kicked out of the boxing academy and the crippling guilt of having so violently thrashed two people, one of whom was Noor. And the fact that I'd done it out of anger, and not a desire for success that cemented my decision to leave. I still remember approaching Ammi-jan that evening and telling her to leave. Ma hadn't done anything, approving of my decision as she scowled at her sister. Pa and Diya had stood, glowering at Ammi-jan, but it was the pure detest on my face that made her leave.

I continue walking forward until a beautiful figure is in my line of sight. I begin to smile and raise my arm in greeting when something uncomfortable crawls up my back. My head whips around to see Ammi-jan standing leaning against her car, having exchanged her kurta for a high-neck, red lace sheath dress. Her sunglasses are pushed into her long bun of black locks, garden hat tossed into the car, and her blue eyes pierce into me. The shrewdness in those eyes as she glances between Kyra and me is enough for me to pull her into the store, blood running cold.

The day's coming to a close. Kyra and I'd shopped, watched a movie, played air hockey and taken a bunch of pictures. Now was finally the time I'd been waiting for. Grabbing Kyra's hand, I drag her into a corner of the first floor, right next to the bathrooms. Keeping a careful eye on the surroundings and the bathroom, I take a deep breath.

'Kyra,' I say. 'I have something to tell you and then ask you.' She tilts her head curiously at me, looking like a cute little puppy and I fight the urge to squeeze her cheeks. 'I know it's been only a little over a month since we've met but I really like you. And I don't mean like a friend.' I pause, staring determinedly at the ground

'I mean I do like you as a friend! But I've realised that I have feelings for you... romantic feelings

'I clarify. 'So wanna date?' I cringe; so smooth Elina, so smooth. Hoping I haven't messed everything, I chance a glance. Kyra's a bright red and the beginnings of a smile appear on my face.

'Oh! I like you too,' she responds eagerly. 'I'll date you.' I chuckle, looking at her pale skin resembling a freaking tomato. 'So, umm... what do we do now?' I gape at her.

'You're asking me? You're the first person I've ever dated I have no idea how this works!' I say. 'Aren't you the expert.' She smacks my arm, face-palming and groaning 'shut up' over and over again.

'Okay then, I'm gonna do something,' I warn. 'So don't smack me too hard if you don't like it.' Grabbing her chin, her wide eyes stare at me as I lower my face. Our eyes flutter shut and I almost miss her mouth. Kyra giggles nervously and then I'm kissing her. Woah, this is so much better than the last one. I pull back after 30 seconds, breathing in large gulps of air.

'That was really nice,' Kyra smiles up at me and I grin. 'Breathe through your nose next time though.' I grin dopily, making a mental reminder.

'What?' She demands, pale skin blushing yet again. 'Why're you grinning like an idiot?'

'Nothing, nothing,' I assure her. 'It's just, there's gonna be a next time.' I can't help but smile again and she chokes on air somehow. Bristling like an offended cat, she stomps away. I watch her go, my eyes dropping to her backside. I whistle. Right, I can do that now. Awesome.

★ ★ ★

'Hey,' Kyra says, snaking her arms around my waist. 'What're you doing?' She kisses the side of my head and I relax against her grip. I gesture at the Spanish homework helplessly; pretty self-explanatory. She snorts and sits down with me on the bed. Glancing around at her room, I raise my eyebrows at the posters of actors, writers, singers and models the walls are covered with. The tall poster of Leonard Cohen makes me grin at the recollection of his voice.

'Your mind better be out of the gutter,' She warns me, eyeing me cautiously. I smirk, she's too easy to tease.

'I was just smiling up at old Leonard. What were you thinking about?' Scoffing, she tells me to focus on my work and shifts onto the desk.

'Here, listen to my pronunciation,' Kyra calls, hand comfortably gripping my wrist after over a month of dating. 'Mon amour est belle.' She repeats it a couple of times until the mistake is evident and she corrects 'armour' to 'amour'. She repeats it, staring directly at me and I blow her a dramatic kiss.

We continue working in comfortable silence until my earlier thought comes back to me.

'Kyra, are you free next weekend?' Kyra frowns. 'I think we need to do something important.'

'Our 1 month anniversary was last week though,' she responds. 'Though I am free, why?' I take a deep breath and gently cup her face between my hands.

'I think we ought to tell Noor and Aditi about us,' I say, voice gentle and reassuring. 'It's gonna be summer break soon and I think they should know.' She remains silent as the words register. And then pure fear.

'Are you crazy?' She demands. 'What will happen if they don't accept us? We need to fit in and be viewed as normal Elina!' I gape, nonplussed at the unexpected reaction.

'But we are normal,' I protest. 'And they'll always be people who won't accept us but we can hardly hide all the time.' Kyra jeers in disbelief.

'Yeah we're normal, but society doesn't think that so we can hardly go announcing ourselves around.' At my protests, she barrels straight on. 'I just want to be normal, to not be an outcast. Can we please respect that?' Sighing, I survey her face carefully.

'Just... just give Noor and Aditi a try. We'll decide the rest later on, okay?' I wait patiently, allowing her time to think.

'Okay,' she relents. 'But we do it next week and we don't tell anyone else no matter what you believe. If rumours spread then we pretend to be straight and normal.' I agree and hug her as thanks. Her grip is uncertain but she clings to me all the same.

We're sitting in a patch of grass right under the shade in the Bio-diversity Park. To our (well mostly mine and Aditi's) chagrin, Noor and Kyra have gotten some textbooks. Kyra bemoans French (as always) and Noor's wailing about Chemistry and History. Aditi and I exchange glances, snigger and tear into our Subways, moaning exaggeratedly over the taste.

'I'll never understand History,' Noor complains. 'It's just memorisation upon memorisation of boring topics. French at least follows a pattern most of the time.' Kyra sniffs in disgust.

'French, predictable? At least I can help you with History but you can't help with French because you're not even learning it!' I cock my eyebrows, bemused by the ridiculous argument going on between the two. A hand slips in mine and Aditi rolls her eyes at me. A small snicker escapes before I can help it only I laugh into the silence.

I look at Kyra to see her staring at our linked hands icily. A strange sense of guilt runs through me before I squash it down. Nonetheless, I drop her hand.

'Guys,' Kyra calls, drawing our attention. 'Elina and I have something to tell you.' She nods at me and I gulp.

'We're dating,' I blurt out and Kyra slaps her forehead. Aditi and Noor gape at us, looking openly between the two.

'You mean you're...' Noor trails off, looking helplessly at me. Aditi looks equally confused beside her.

'I'm bicurious, though I'm getting convinced of my bisexuality day by day. Kyra is,' I squeeze her shoulder encouragingly. 'She's bisexual.' Noor seems to be calming down and beams at us while Aditi still seems a bit shocked.

'How long?' Noor demands.

'A month.'

'Do you like each other?'

'Yes.' 'Yup.'

'Then I approve,' Noor allows.

'Gee thanks,' I laugh, feeling as though part of the humongous weight has been lifted away. Glancing over, Kyra seems to be smiling gleefully as well. Aditi clears her throat and looks at us intently.

'I fully support you and thanks for trusting us,' she says seriously. 'Also I... um, nothing... nevermind.' I wait to hear if there's more but let Aditi take her own time. She doesn't meet my eyes though, expression twisting into guilt, so frowning I decide to leave the matter for another day. Kyra's watching her as well, and it's as though she's beginning to understand something.

'So what're you guys doing now?' Noor asks and shrugging, my girlfriend tells her we'll probably go on a date. Aditi wolf-whistles as we bid the two goodbye and Noor extracts a promise of a double date with her and her boyfriend.

Booking an Uber, we decide to go visit my boxing academy. Even if I can't go inside, I want to show Kyra how it looks. We pull up on Bahadur Shah Zafar Road with the academy looming in front of us. The entire structure is perhaps the size of the school's concert hall and bleached white. A humongous gate guards the entrance with a fenced wall running around the structure. At present, the gate's closed and the doors are shut. A sharp pain seems to pierce through my chest at the sight of the building which had once formed my dream being closed off to me.

'Hey,' Kyra says, squeezing a cheek. 'None of that miserable look, okay? I think you should try boxing again.' I stare at her in surprise and she shrugs.

'I think you should give it another go. What happened with your aunt wasn't your fault and you need to move on,' she continues. 'During our break, come back to this academy and explain things to your old coach. If that doesn't work out then join a new academy. Don't just give up.' I exhale, knowing that she's right but that it hurts more than I'd like to admit.

'You're awesome y' know,' I mumble, squeezing her hand warmly. She chuckles; 'I do know.'

It's nearing the end of our break, July 20th, when I muster up the courage to walk into the academy again. I'd contacted Coach in advance, explaining my situation as truthfully and explicitly as I was comfortable with. He'd allowed me to talk and then listening to my experience in therapy, agreed to give me a second chance. I look around, taking in the equipment, the rings and the punching bags. It feels good to be here but damn am I nervous.

'Elina Dutta,' Coach calls, standing inside the first ring and towering over me. 'You're exactly on time.' Coach looks as enormous as he always does, standing at a cool 6'2 with his chestnut skin and toned physique. Amused blackish-brown orbs glint at from behind thin, steel glasses.

'Coach Singh,' I grin. 'It's been a long time sir. New look?' As usual, he's wearing shorts and a tee, showing off his bulging muscles. The unusual is the new classic goatee he's sporting. He bares his teeth in response and calls me on the ring.

Hoisting myself up, I step foot inside and close my eyes for a few seconds. Before I can dwell in my sentiments Coach speaks a 'Think fast!' and throws a glove at me. Narrowly dodging it, I grab the offender and put it on. The same repeats and I'm soon geared up. The only piece that's missing is the headgear.

'You stay in the ring with me for 5 minutes and you're in,' Coach says mischievously. 'Without the headgear of course.' The bell rings and we begin. I'm on the defensive; with Coach, the best idea is to defend, block and dodge attacks and this works out perfectly for a set time limit. I fall into a familiar pattern of repeatedly dodging and blocking Coach's hits until a hit lands on the side of my head. The struck area throbs and my vision swims. The strength is carefully controlled but it's hard enough for me to land on my ass on the next shot. And looking at Coach's smug little grin, that was the whole point. The old strategy needs to be distracted. On the offensive now.

I circle him, protecting my face and my ribs. Jumping forward, I jab at his chest and then leap back, narrowly missing the hit to my ribs. Feinting, I leave my stomach unguarded and he strikes. I draw my arm up at the last second and block the hit, wincing as the vibrations travel through the numb area. His ribs are unguarded. Shooting my right hand forward, I nail his ribs and strike my other fist against his chin. He stumbles and then straightens, beginning to advance and then the bell rings. The 5 minutes are over.

Gasping, I slump to the ground and accept the offered water bottle gratefully. Coach Singh grins at me, pressing an ice pack against the throbbing area.

'I'm still expecting a formal apology for Fatimah,' He warns. Right, that's the name of the girl who I beat up. 'And for Noor even if she's your friend and you probably already sorted it out.' I nod, still in a daze.

I have another chance. The ride home seems surreal until I find Kyra eagerly waiting for me in my room.

'How was it?' She asks, shaking my arm eagerly. Immediately her expression turns alarmed and she pulls me into a hug. 'Hey, hey don't cry love. It's okay.' Wordless, I trace my cheeks and sure enough, they're wet.

'No you don't understand,' I say, mumbling into her shoulder as I grip her tighter around the waist. 'I'm in. And it's all thanks to you.' She shrieks in excitement and hugs me fiercely. She's rambling about something but all I can think about is how much she means to me.

Gently, she presses our foreheads together and I focus on harmonizing our breaths. Leaning forward, she pecks the corner of my mouth and wipes the tears away. And the door slams open. I forgot to lock it.

Ammi-jan stands in the doorway, a look of utter terror settling on her face.

'Get the hell out!' She screeches at Kyra who draws back in horror. 'Get out you little slut, you freak how dare you even step foot in here—'

'Don't you speak to her that way!' I scream back, shoving Kyra behind me. 'This is none of your business so stay out!' Her eyes seem to go right past as she lunges forward, snarling inhumanely at Kyra and my vision blacks out.

A moment later it returns and I've pushed Ammi-jan into a wall. She flails against me and finally kicks me. Immediately something snaps and I knee her in the stomach. Startled shouts sound beside me and Pa pulls me back. My gaze remains firmly trained on the gasping form of my aunt, though I keep a careful eye on Kyra who's now being shielded by Diya.

'What happened here?' Ma demands, looking at me, Kyra and then her sister. 'Does someone want to explain to me why we heard so much shouting and then walked in on Elina kicking my sister?' Ammi-jan gasps and then attempts to sit up. Ma rushes and props her up.

'I'll tell you what happened,' she pants out. 'I walked into the sight of the disgusting freak kissing Elina. And the moment I tried to get it to get out, your daughter went crazy and attacked me.' The stillness after this has Ma, Pa and Diya trying to understand the implications. Their shocked expressions ignite a rather desperate need for reassurance.

'You asked me if I was inspired by anything when I drew that painting Pa,' I say, looking desperately at my father. 'I was inspired by Kyra.' Pa blinks, thrown off-kilter by the unexpected response. Ma and Diya stare at me in contemplation and experiencing a sudden burst of courage decide to lay it all out.

'Kyra and I are dating.' I grab her hand. 'I'm pretty sure I'm bisexual.'

After a few agonising moments, Diya scoffs; 'Well that was dramatic! If darling Ammi-jan wouldn't have interfered then Elina could've comfortably told us in her own damn time.' The tension suddenly breaks and I grin at my sisters' response.

Pa nods, smiling at me; 'The community has always had my acceptance and respect. I'll treat Kyra like how I might treat any potential boyfriend of yours.' Giggling, I look over hopefully to Ma, only to falter at the saddened expression on her face.

'If I could be brutally honest, then I'm 20% disappointed that you're bisexual and might not marry a man,' she says as Ammi-jan chuckles. 'I never wanted my children to belong to the community but if you're happy then I can set my disappointment aside.' A burst of fondness spreads throughout my body for the family who accepts me. I'm truly lucky and very grateful.

'And you're just fine with all of this unnatural behaviour?' Ammi-jan sneers, glaring at us. 'No one would mind if I told the girl's family, right?' Kyra splutters and begins protesting from beside me.

'Enough!' Ma snaps, scowling toward her sister. 'Geeta you have no right to try and dictate Elina's life nor expose a child's' secret to their parents. Stay within your limits.' The woman who was once my beloved aunt stares at me in repulsion.

'If you want someone sick in your house then you can keep her,' she dismisses. 'Besides, this abnormal will probably think of her sister in the same perverted way soon enough. Sick creatures!' That's the last straw and finally, I snap altogether. Jerking forward, I begin pushing Geeta out of the house and out the door. Eventually, I hope to forget about her altogether. Finally left alone again, I can sense the apprehension pouring out of Kyra.

'She can't tell my parents Elina!' She screams, eyes wild and frightened like never seen before. 'My parents would never accept this they'd hate me!' I place my hands on her cheeks attempting to calm her down when she slaps them away. I withdraw, hand stinging and a bit bewildered.

'They won't know until on our terms Kyra,' I attempt to soothe her. She continues to ramble helplessly, refusing my aid and ignoring everything that I say. That's the time we came undone.

Kyra's grown distant since the beginning of August, hiding behind various masks when she lies. And she does so all the time. She lies about her feelings, about us and where we stand, everything. I want to make amends so I seek her out, racing down the school corridors and attempting to find her. A passing teacher tells me she's in the gym.

Grinning I walk in to see Kyra locking lips with a guy I've never seen before in my life. They break apart the moment I barge in and I can see the horror and guilt as clear as day on Kyra's face. I can see so many emotions but not regret because she doesn't regret it, not really.

'I'm sorry Elina,' She whispers as the guy nervously chuckles and walks out. 'But my parents want me to be normal.' And of course, they do. I leap forward then, my hands wrapped around her throat and squeezing just a tad bit more— but no, that'd be too dramatic. She's gasping and panting for breath and so I drop her on the ground. I see everything and I walk out from the beginnings of a cry.

Because I can wear a mask as well. Now whether it's a mask of kindness or cruelty I'm yet to decide. Kyra's parents get a text message; it's a picture of Kyra locking lips with a girl at her previous school. Social media and bad break-ups are good sources, easily willing to.

It's on my Insta story, circulating for almost the whole of school to see, either by reposting or by following me. The mask of normalcy she wears has been shattered. Question is, which one will she put on next?

Only she doesn't. Kyra transfers out of St.Alexius and moves back to Lucknow. I only know of this after running into her whilst she returns from the principals' office without her ID card. She opens her mouth to speak but I move forward.

How easily my anger can turn deep-rooted fondness into deep-rooted bitterness. Perhaps that's my most dangerous mask, the one where I pretend I don't have one. For the first time, my words and my silence has hurt someone more than punches.

Aditi walks toward me, the day after the transfer, newly dyed blue hair billowing behind as she stormed up in her football jacket and slaps me in the cafeteria. She leans toward, mean and grim;

'If I'd told you that I was pansexual, is this what you'd have done to expose my vulnerabilities?' She sweeps away like the storm she entered as. And it's that realisation that maybe, I broke my friend's trust is what hurts. But at that moment, I'm a burning red flame and my anger will destroy everything and everyone around me. And that anger is the most dangerous mask of them all.

Anger-III

Acceptance

I'd admitted to both adore and despise Kyra. A fact considering it's the same for myself. I'll reveal now that she is the one who made me write this diary after all.

October's coming to an end with brown leaves falling all around me as I walk through the streets of Chandni Chowk, excitedly chattering on the phone with Noor about my applications when I see her through the crowd and I know she sees me too.

It's always made me wonder if she hadn't been so afraid to show me her true self and hadn't hidden constantly behind a facade, would we have been hurt as terribly as we were? Or perhaps, if I had not pretended to be without flaws, had understood the necessity of my masks would I have been as hurt as I had been? It's perhaps with that knowledge that I tell Noor that'll I call her later and hang up.

The once-exhausted smile comes easily after three years of separation. We're both wearing light sweaters, her white to my black, and where she wears leggings I wear trousers. Her hair is open while mine is caught up in a low ponytail. It takes a few moments to note this description and it's obvious she's surveyed me as well. We are similar yet different, in a lot of ways. And we have a lot to talk about.

I smile and beckon Kyra over.

We're sitting in a Cafè Coffee Day shop, I realize belatedly, taking note of the sweet scent of roasting coffee beans, the high rising tables and the busy countertops. Kyra's sitting across me, hands cupped loosely around her steaming cup, illuminated in the fading light of the sunset as she takes tentative sips, ignoring the apparent heat.

We've talked for hours about insignificant topics, make small talk as though to lessen the ice wall formed around us. I stare down at my cup, peering absentmindedly at the heart-shaped cream coating the surface, and wonder what made Kyra agree to follow me.

'Perhaps,' I think, 'She too believes this conversation is long overdue.'

She clears her throat suddenly; a soft, attention-grabbing noise causing me to violently jerk my head up in surprise.

"So, New York University Tisch School of Arts was it?" Kyra says, smiling slightly. She smiles a lot less than she used to, I realize, but the few ones she produces are genuine.

"Yeah," I reply, smiling back. "And then after three years, it'll be off to The Julliard School which is also in New York. Boxing is something I'll still be pursuing but I can't make that last forever or make much money out of it. And you're aiming for St. Xavier's College. Not gonna lie but I didn't think you'd ever want to be a Professor of History."

She chuckles along with me, a shared mental image of Kyra teaching an almost dozing group of brats about the times of Akbar or the ways of Napoleon in a stuffy classroom springing to mind. Her other passion of becoming a museum conservator doesn't seem half-bad either.

"And how're Noor and Aditi?" She continues in an obvious attempt to not let the conversation die out.

"Good," I reply, slightly awkward about the standard response. "Noor's going to Harvard to get a P.H.D in Law; she's finally decided to put off dating for a while to focus on her career but I doubt her ability to stay abstinent for really long." We share an amused smirk.

"Aditi's still pursuing football, pretty sure she's going to force India to participate in FIFA at some point..."

The conversation dwindles, occasionally interrupted by the chiming of the bell as the door to the Cafe is pulled open or shut.

"She's pansexual, she told me after you left," I continue. "How is your dating life by the way?"

Kyra looks around uncomfortably before responding that she's not going to date for a while before asking me the same.

"Dated a few guys, then a few girls," I shrug. "Not very interested in anyone at the moment, haven't been for some time."

We descend back into silence until I just can't take it anymore.

"It took me a while after you transferred," My voice is soft, small. "I pretended those months didn't happen and just ignored everything except for my studies. It took me a vague decision with P—My father to realize how toxic we'd been." I look at her, watching the expressions flit across her face.

"I don't believe that I can blame you for everything, we were both just kids," She looks guilty now with her downcast expression. "I was too caught up in this world of rainbows and sunshine where everybody accepted us.

And when I saw you kiss that other guy, my anger just went out of control. What I did after the breakup was not okay—"

My chest heaves with a large breath at the recollection of my hands squeezing her neck and maybe just a little more.

"I'm sorry for that," I finish out. "And I hope we can move on."

She stays silent for a long time, dark lashes lowering over tightly screwed shut eyes as though attempting to block out everything around her and think.

'My parents nearly disowned me after that stunt,' she says, eyes still closed as she completely catches me off guard. 'It took my grandparents and cousins talking them out of it. I've been in therapy for nearly a year now. Everything that I was upset about, angry about, it just felt good to let it all out and try to move on. To realize that it's okay to feel pain no matter how insignificant, to learn to embrace and to let go. It wasn't easy to realise that I didn't have to base myself of what someone considered to be normal, especially not when it brought me grief. And for hurting you, I'm sorry.'

Her voice nearly breaks towards the end as she opens watery eyes.

"I shouldn't have been so afraid and ashamed to show you my true self, maybe I should've accepted myself before I tried to accept you," Kyra continues, fiddling with her sweater sleeve as she maintains my gaze.

"I'm sorry."

A smile breaks out after I tilt my head toward the sleeves, an old gesture inciting many memories. She flicks the sleeve back and her skin is clear, smooth and so pretty.

We're both sorry and we're both striving to move on. It's like a weight's been lifted off my shoulders; that nagging feeling in my mind about something that was missing is finally gone. And judging from the obvious slump in her posture, Kyra feels the same way.

"I did like you, you know," I say, surprised by how easy it to admit it now after this talk as compared to after the break-up."

She laughs openly, aiming joyous eyes at me;
"I know, I liked you too."

We exit the Cafe together and I hug her tightly, burrowing my head in her neck and squeezing her arms until we've parted ways.

Standing on the ring again, it's not easy to move on. It's not easy to realise when I need to accept myself and show my true self around others, and when I need to hide. Diya whistles as Ma and Pa cheer. Aditi and Noor wave about banners as I realize I'll always wear half a mask burning with anger. And that's okay. I throw my first punch.

'The love that dare not speak its name... It is beautiful, it is fine, it is the noblest form of affection. There is nothing unnatural about it.... That it should be so, the world does not understand. The world mocks at it, and sometimes puts one in the pillory for it.' ~ Oscar Wilde.

Denial

– Sadhika Anand

Denial-I

I wake up to the loud ringing of my alarm. I turn it off but I don't get up from my bed just yet. I just lay on my bed, thinking and overthinking about anything and everything. It's become a habit now. Time went by really fast when I did this earlier but the last few months have been different. Ten minutes feel like an hour and an hour feels like a day but I refuse to give this time up. I need some parts of my life to remain the same even when my whole world is falling apart.

Who am I? How was the Universe created? Is there life after death?

After months of wondering, I believe who I am is a question that I will probably never be able to answer — I am a different person every day. And yes, I do believe that there is life after death. I'm not sure whether I believe in heaven and hell and Satan and God but I do believe that something exists in this universe, something that we'll probably never be able to figure out. We can't just be gone forever, can we? There must be someplace people go after their heart stops and they take their last breath. And I'll be able to tell you how the universe was created after I can figure out whether or not something existed before it because if there was nothing, then what lead to the creation of the universe? Surely it wasn't just a loud bang and boom there it is the Universe.

But today, I can't concentrate on anything. Today all I can think about is her.

It's been three months since I last saw her or spoke to her. I miss her. I miss her and I miss all the things we did together. I miss going to school with her, I miss our sleepovers, our talks, our discussions, everything. I miss the comfort and safety of her home and the way she managed to light up every room she walked into. Most of all, I miss seeing her face every day – and there's nothing I can do about it. I can't text or call her, I doubt there's network in heaven, or wherever she is.

Dead. No matter what the conversation is, they'll always talk or at least think about how she's no more. Personally, not being able to grasp her absence, the fact that they mentioned it in regular conversation was astonishing to me. There are moments when her being gone completely escapes my mind. I'll be sitting on my desk, studying and I'll think of something and the first thought that would come to my mind would be "I have to tell her". And when I pick up my phone to do so it hits me. It hits me on the head like a 1concrete block that she is gone.

I get up from my bed before my thoughts make me late for school again. I've barely made it on time this past week and if I am late today, I don't know what'll happen. As I get ready I wonder; How long will it take for me to finally forget her? How much time has to pass by before

I finally stop feeling the pain? How many months have to go by before I can finally accept that my best friend is gone forever?

I don't think there will ever be a time where I don't remember her, don't miss her. After a few years, maybe I won't feel the pain in my chest that I feel now, every time I hear her name, every time someone mentions her. Maybe I'll forget how she looked or what her voice sounded like. But I don't think there will ever be a time in my life where I don't remember her or miss her. There will never be a time where I forget her and the time that we spent together and the memories that we made. And I know, wherever she is, she won't forget either. After all, how do you forget something as beautiful as what we had?

I finally gather some will power and convince my mind from thinking whatever it is thinking. As I finish getting ready I look at my reflection in the mirror. The skirt feels tight over my thighs and no matter how many times I tug at it, it keeps riding up, exposing my rather well-fleshed legs. My stomach is protruding from my shirt as its short sleeves barely cover up my drooping arms. It takes everything in me to not tear away these clothes from my bulky body and wear my pyjamas and stay home. I stare at myself in the mirror still not able to understand how I manage to show up at school every day when I look like this.

I get properly dressed and leave for school. As I walk down the corridor, constantly pulling down my skirt and adjusting my shirt, I notice everyone looking at me. I always do. I see the stares and the laughs and the weird looks that are passed when I walk by. I tell myself that it's nothing but deep down I know that it's not. I try to ignore them as much as I can but some days it becomes extremely difficult to do so. I always knew my body looked different than my friend's bodies. But I never thought that the difference would turn out to be such a bad thing that I'd lose friends over it.

But not her.

She was the only one who decided to stick around even after everyone else thought that I was too ugly to be talking to them, the only one who believed that I was worthy of something beautiful and deserved love and happiness. We were each other's only comfort. She used to tell me all her deepest, darkest secrets and thoughts and I did the same with her. Before her, I used to keep everything bottled up inside me. I was afraid to admit my feelings to myself and I was terrified of the thoughts in my brain. But after she started talking to me I felt as if I finally had someone I can share my feelings with. It took me a while before I opened up with her but once that happened, there was no going back. She confided in me and I confided in her. She told me about her old school and her old friends and bullying and anxiety. I told her about my family, my body, my writing and my thoughts. She was the only one I had to call whenever something was bothering me and I was the only one she called. We were as comfortable with each other as someone is with their blanket on a cold winter morning. When I'm in my room crying because my heart hurts, the only thing that stops my tears is the thought that wherever she is, she doesn't have anyone either.

Now that she's gone, I walk the school corridors alone – but I have to be strong for her. I know that she's looking, wherever she is so I hold back my tears and keep my chin up, for her.

I started writing a diary. I named it after her so that it made me feel as if I was talking to her. This didn't help as much with the pain as I thought it would but I felt relieved as if she wasn't gone. I was comforted by the fact that she was still with me and I could talk to tell her about my day, even if she couldn't tell me about hers.

I reach my class, panting and out of breath because my small heart cannot tolerate the weight of my bulky body. I sit down at my usual desk, the last desk at the corner of the classroom. The entire day goes by just like it does, teachers coming in and out and the students making the most of their school life. People talk to their friends, telling them about their weekend while I sit at my desk, with my nose in a book as usual, in another universe entirely.

The last bell of the day rings and I finally feel free. I pick up my bag and as I exit the prison I call the school, it is as if a huge weight has been lifted off my back. I feel this way every single day until the next morning when that sinking feeling comes back. I hate school, in case it wasn't already obvious. I don't hate waking up early and getting ready part of it as most people do. It's not even the studying that makes me hate it. What does make me hate it though is how it makes me feel- incompetent, inferior.

Up until the age of twelve, I was an ideal student but then everything went down the drain. I still get good grades but hitting your pre-teens makes you lose all your confidence. Every single time I stepped on stage, a voice inside my head told me that everyone in the audience was judging me. They weren't judging my speech, they were judging how I looked. The first few times, I ignored that voice and carried on but after that I just let it take over me. that voice now decided everything I do and how I do whatever it is that I'm doing. It's as if I don't have a mind of my own anymore. That voice dictates me and my every action. I hate it, yes I do. But I don't dare to do anything about it. I want to stand up against it but I have never been able to, even when she was there right next to me.

I take a detour on my way home. Instead of taking a left turn and going home, I turned right and see her house and it is as if a movie starts playing in my head and that day flashes across my eyes.

I was sitting at my usual desk, the last bench of the corner row. The seat next to me was empty today and I had no idea why. I spoke to her last night and she told me that I'll be seeing her today. I was confused as to why she would skip school after telling me that she'll come. It was wrong and unfair to me. She knows how uncomfortable I get when she's not with me. She knows all of it and yet she missed school without telling me. I'm worried about her and I have no one to talk to. The least she could've done was text me and let me know that she was okay but she'll be missing school. A school without your best friend there is no fun at all.

Suddenly there was an announcement over the speakers. We were all supposed to quietly go down and assemble in the auditorium. So that's what everyone did, except no one was quiet and instead of assembling in the auditorium everyone was talking to each other about the football game last night or the latest episode of some TV show or what dress they'll be wearing to the party on Saturday. I think I was the only one who did what was asked. But all conversation died when the Principal took to the stage to address us all. Initially, I thought he'd say something about our school basketball team winning the state tournament but little did I know that what he was about to say would change my life forever.

When he took her name, I immediately knew something was wrong and when he said the word 'died' in the same sentence, my heart broke into a million little pieces. All air left me and I lost all control over my body. My legs were shaking; my hands were trembling. As tears rolled down my face, all my mind could manage to think was that this cannot be happening. No way was she gone, forever. A loud noise erupted around me, everyone talking about her and what could've happened. All we were told was that she was no more. Before I know it I'm standing up along with the other students and going back to my class. My feet are being controlled by my mind but the mind doesn't seem to be my own. We're being excused early today. I pick up my backpack and run. I run as fast as my feet can run. By the time I reach her house, I'm panting. This time it feels as if the air has exited my body and I'm left begging for every breath. Is this how she felt before she died? Maybe she did. My mind was unable to comprehend what was happening. One moment I was cursing her for not coming to school and the other I was burdened with the news of her death.

reason for so many of our smiles. I can see her mother sitting next to her, too numb to eve"I was burdened with the news of her death". I repeat the sentence again and again in my mind. The words 'her death' seem unreal, as if they're not true and all of this is just a dream. It feels like a dream. There's a hole in my heart where she used to be and the pain is unbearable. I can't wrap my head around what just happened.

The verandah where we spent all our afternoons is filled with a mass of sobbing people I don't recognize. I didn't notice when the tears started falling but I can taste the salty water on my lips as I entered her house. There she is lying on the floor with white covering her body.

It looks as if she's sleeping and all I want to do is go to her and wake her up. But my shaking legs stop me from moving even an inch. People around me are crying, mourning the loss of this beautiful creature who was then cry. She was caressing her hair as if putting her to sleep. When she saw me she frantically started calling out my name. she wanted me to come and wake her daughter up for she knew that if there was anyone in the world who loved her half as much as she did, it was me. I went and sat down right next to her, for once not caring about what my body looked like.

After the last rites, everyone was supposed to go home but I stayed there. I stayed at her house, the place where we spent every minute of every day. I spent the rest of the week there, with her mother. Together, our grief felt like comfort and even though pain wrapped around us like a cocoon, we became each other's support. But she couldn't stand it. She couldn't stand living in that house alone so she moved away making me feel once again, alone and abandoned.

The police told us it was suicide. Her anxiety got the best of her and instead of calling me during her panic attacks like she normally did, she decided to end it once and for all. I thought she told me everything but she never told me she had suicidal thoughts. And according to her therapist, she never told her either.

For a very long time, I blamed myself. I should've noticed it; I should've known that something was different. But I was too self-absorbed, too focused on my problems to notice what was going on with her. sometimes I still think that it's my fault but other times I tell myself that there's nothing more I could've done except provide her with the support that I already was.

I look at the building in front of me and breakdown into tears, right in the middle of the street. Where once her house stood, now stands a brand new house. Even though every part of the old building was rebuilt into this new structure, I still feel her presence whenever I come here, almost as if she stayed here even after her mother left. As if she stayed here for me. I go to the park in front of the building to look at the flower beds. 'Planted in memory of Naisha Gupta', it said on a board in front of a bed of pink peonies- her favourite flower in her favourite colour. Her name meant special and oh was she special. Her smile lit up every room, and her voice was like honey. She was beautiful and boy was she special.

I stayed for a while and then walked away after taking one last look at the house I once used to call home. The fact that she was gone is still not something I have been able to accept but one day hopefully I will be able to. All I know is that the day she ended her life was the day all colour washed away from mine. She did not only take her life, but she also took my happiness. And I know that no matter where I go in life, no matter what happens, I will never be able to be truly happy again the way I was. I will never get to experience euphoria; I'll never be in a state of complete happiness and bliss. There will always be emptiness in my heart, a void left unfilled because of her. There will always be an abyss of grief inside of me that no amount of diary writing or visiting her house would lessen. I might not have accepted her death and life without her. But what I have accepted is that I will never be happy again. I will never be able to accept that there is still happiness left for me to find. She was the reason, the source of the smile on my face and without her I can't help but think that I will never truly be happy, no matter how many facades I put up.

Denial-II

The rest of my day goes by just as normally as it would every day. I finished my homework, got my daily dosage of the family drama and also managed to feel hopeless about my life. But while I was going about my day the only thing my eyes could see was her face and the only thought that came to mind was her death. It didn't matter that a psychology book was staring at me, for even in Wilhelm Wundt's face the only thing I could see was her.

History and politics and psychology- We often talked about these things. A lot of our discussions consisted of her talking about how one day she'd like to take over the world and make it compulsory for schools to teach students about the real things in life and not about mitochondria and trigonometry. She wanted that students be taught about taxes and mental health and global warming. She once researched what students actually wanted to learn and then made a petition for it. A lot of people signed it but it never really managed to bring about a change. She used to make and sign a lot of petitions. It was her way of saving the world. That's one thing she was passionate about- saving the world; not from aliens and monsters but humans. She believed that we were the real monsters and that one day it will be us that cause the world to end and not an alien invasion. And I agree with her. Humans will be the only reason for the human downfall.

My psychology book must be thought-provoking because just as I start reading about the evolution of psychology, I can't help but think about the evolution of humankind. How did we start with nothing and end up here? What was the one thing that brought us where we are today? What was it that led to us living in a world of robotic servants where everything is available to us at the click of a button? Was it simply the human mind-constantly evolving and developing? Surely it was our brain but there has to be something else too. There is no way it was just electrical signals passing through our bodies. There has got to be something more.

And just like that my brain takes off and I start my journey to the land of unanswered questions. I'm always amazed when this happens. It never fails to shock me how my brain just starts thinking about such weird questions simple after looking at something completely random. I think it happens to everybody, or at least I hope it does. It's the most wonderful feeling in the world; asking questions to yourself that no one really knows the answer to and then coming up with your conclusions and beliefs. I hope that everyone experiences it, that eureka moment when you finally found the answer that you've been looking for, at least once in their lifetime. The human brain and robots and evolution drift me off to sleep.

I'm standing by a lakeside, stagnant but deep. The water is as clear as a crystal, so much so that I can even see the rocks and the soil beneath it. I'm surrounded by nothing but emptiness and a lot of greenery. There's a voice in my head telling me that I know this place even if I don't recognize it right now. I try to think why I'm here, my mind confused and foggy. I know there's something special about this it but no amount of pushing is helping my brain remember.

As I'm looking around, trying to find something of significance, I see something move across the lake. There's something, no, someone coming from inside the woods. My entire body freezes when I finally see who it is. My eyes start welling up and chest tightens. It's her. It is Naisha. My eyes don't believe what they're seeing just as my brain can't comprehend it. It takes me several minutes to process the fact that I'm seeing her, really seeing her.

Her face looks exactly how I remember and her hair was the perfect shade of black as usual. Her skin looks pale and yellow but she looks beautiful. I think in my head that she looks like herself, she looks like Naisha. But then her eyes meet mine and I feel as if I don't recognize the person I'm looking at. Her eyes aren't twinkling as they used to and she doesn't look happy. Instead, she looks sad. She looks like she has been absorbed by sorrow, just as I had been absorbed by grief. I recognize her and yet I don't.

Nonetheless, I want to swim across the lake and go to her and hug her and tell her how much I miss her. But I cannot move my body at all. My legs are frozen and no matter how much I try, I can't move. Right now, at this moment, there is nothing I'd rather do than talk to her but my mouth is as frozen as my legs. I can't speak a word, let alone tell her how much I miss her.

She looks at me, her eyes staring into mine from across the lake, but she doesn't say anything. She moves her glare from me to the lake and looks inside. The clear water lets her see everything, the soil and the rocks, the fishes and the plants. It confuses me but somehow I manage to form words and open my mouth, "Naisha? Naisha?" I say but she doesn't respond. "What are you doing?" I ask her but she goes on staring at the lake. "I miss you. I miss you so much." And to that, she looks up but then quickly shifts her eyes back to the lake. Just as I'm about to say something else, she jumps into the water and a right light blinds me. I hear the splash in the water as her body touches its surface and before I can process it, she's gone.

I wake up in my bed, my head throbbing because of a headache. I open my eyes and everything around me starts spinning in circles. I feel dizzy as if someone hit me with a bat and then left me here, passed out. A wave of nausea hits me just as I attempt to get up and start my day. I try to think of something, one of my questions that I ponder about this time of the morning but nothing comes to me and when it does, my mind just cannot think about anything. There are a million voices inside my head shouting a million different things but I can't focus and listen to just one of them. My stomach churns just as I'm reminded of my dream last night.

I saw her. It was the first time she came to me in my dreams ever since her death. Her mother saw her, her cousin saw her and even a few of her friends from the school said that she came to them in their dreams but it didn't happen for me. I had given up all hope, convinced myself that it won't happen but then it did. I wasn't expecting it at all, especially not like this.

The dream was confusing and I'm not sure what it means. Why was I at the lake? And why did I have the feeling that it had some connection to my life? Why was Naisha staring at the lake as if her life depended on it? I guess these questions will just become a part of my long list of unanswered questions that I'll have to think about every morning.

"Kiya" my mom shouts from the living room. "Get up! It's already 10. You have a class at 11:30!"

My mother- the kindest, strongest and most wonderful woman living on this planet. She's my rock, she's everyone's rock. Everything I am and everything I aspire to be, I owe it all to her. If it wasn't for her, I don't know what I would've done with my life. She keeps me sane even when the world around me is going haywire. If there was one reason, one person that helped me deal with Naisha's death, it was her.

I have a crazy family. My dad, my mom and my grandparents. My grandparents don't live with us but they're a big part of my life. We fight a lot but at the end of the day there's nowhere I'd rather be than with them. They are my whole world and no matter what, my life revolves around them. Sure we have our problems. We fight, we argue, no one understands each other here but when it comes to it, all of us would do anything for each other in a heartbeat. That's what family is, right?

"We have lunch at Dadi's place today. Please don't be late, okay?" my mom says as she puts my breakfast in front of me. "I won't" comes out of my mouth just as the first bite of my sandwich goes inside it. A very simple response on the outside but the inside, I'm completely dreading it.

I'm not sure how but I completely forgot about this lunch. How could it skip my mind? Mom told me about it two weeks prior because that's how long it takes me to prepare myself for lunches and dinners and breakfasts and teas at my grandmother's. It takes everything in me to step out of the house to go meet her and it takes even more strength for me to step inside her place and meet her. I know she's my grandmother but I can't sit down with her and talk to her because every conversation ends up with her telling me I should lose weight and me feeling miserable about myself. I love her. I love her a lot but sometimes I just wish she understood me more. I wish she was a little more sensitive to my feelings. Whenever I meet her, I always leave feeling miserable and angry. As if my worth can only be determined by my weight. The lesser the number is on the scale; higher are my chances of being worthy. I mean shouldn't it be the other way round? My worth should go up just as my weight increases? That's the way it is with gold and other precious things. Am I less precious then?

My grandparent's house is lavish and beautiful and expensive and everything inside her house is also lavish and beautiful and expensive. The driveway is lined with beautiful plants and the porch has a beautiful statue of a dog. Inside there are several rooms, so many that one could only wonder what an old couple living alone would do with all of them.

I am a large girl but whenever I'm here, I feel very small.

After the greetings and the usual small talk, we all sit down to eat. As conversation ensues about the big things and important things I can't help but wonder would my relationship with my grandmother be different if I looked different? Or would she have then found something else to taunt me about? I think it'd be different. I always feel this way. I always think that my life would be different if the number on the weighing scale was lesser. It might or might not be true but I truly believe that my life would be much happier if it wasn't for my abnormally large body.

"Kiya", my grandmother's voice brings me back to the real world. "How's school going?"

"Going well", I reply. I try to keep my responses short when it comes to talking to my grandmother. It helps me avoid any further scrutiny about literally anything else in the world.

"What are you doing for your weight?"

"Nothing"

"Nothing? You're still doing nothing? It's shameful how fat you're getting but you're doing nothing about it."

The words leave her mouth and tears start forming in my eyes. "Shameful" was the word she used to describe me. How is it shameful to be fat? Somehow I control my tears but the rage and humiliation that I feel right now, controlling it is out of my hands. I'm about to say something but then I see my mother looking at me from the across the table with an apologetic look in her eyes and I swallow my words back, hoping they would fulfil my hunger.

When I reach home, I go straight to my room. I stand in front of the mirror, looking at my reflection and without any warning, tears start rolling down my face. I cry and I cry, endlessly. Within a few minutes, my face is red and blotchy and my eyes are swollen. I feel disgusted, disgusted with myself. I never did but today my grandmother made me feel disgusting. Is it shameful to be fat? Should I be ashamed of myself for looking the way I do? Should I hate my body and feel less because of it? My answer to this has always been no but today, that no is leaning towards a yes.

As I always do when I'm upset or happy or angry or excited, I take out my diary and I write. It feels as if I'm talking to Naisha. Everything written in this diary is things I would tell her which makes it very confidential. Writing makes me feel better. It makes me think about what I'm feeling instead of just merely feeling it.

As I start to write, the words automatically form. I don't have to think about the grammar or the spelling, whatever I'm feeling translates onto the paper itself and before I know it I've written a poem.

"I walk down the school corridor

Pulling down my skirt to cover my bulky thighs,

I adjust the sleeves of my shirt to hide the marks on my arm.

The marks that scream and tell you the amount of space I take up in this world.

The marks that tell you how stretched and distorted my body looks.

Everywhere I go, I find eyes scanning me.

My hands, my legs, my stomach, my fat.

I see the stares and the laughs.

I see how they look at me,

As if I'm from Mars.

The extra fat you see around me is not as simple as a byproduct of me stuffing my mouth with junk.

That extra fat you see around me is not as simple as me being too lazy to work out.

That extra layer you see around me is me.

It's the humour, the charm, the intellect.

It's everything you like about me.

It's everything that I'm made of, everything that I can and have accomplished.

And how is it my fault if everything that I am was too much for a tiny body?

Why is a bad thing that all the good things about me are present in extra large quantities?

Why is it then, that I am being punished for simply being myself?"

The second my pen stops writing, I let out a breath I didn't know I was holding. It feels good to write it all down-liberating. Writing always makes me feel that way. Everything I feel inside me flows out through the ink of the pen. Even when I was a kid, I used to write a lot. In a way, words have always been my best friend even if only on paper. It takes a lot for me to gather the strength to pick up the pen but when I do I feel something let go in my chest; as if my soul feels free.

My soul. That dream.

And just like that realisation struck me. That dream I had earlier does have a connection with my life. It wasn't just a random glimpse of some random river. It was The Blue Lake in New Zealand and it had a connection to one of Naisha's great plans.

The Blue Lake is the clearest lake in the world. So clear in fact, that you can see through it. The plants, fish, rocks, everything inside of it can be seen clearly. In one of our discussions, Naisha talked about how the biggest superpower one can have is to be able to look inside someone's soul. She had a theory about what made everyone unique and different from others and the soul staring would help her determine that. She believed that every individual on this planet has the same qualities. We're all kind and funny and a little arrogant. Then what is it that makes us all different? Her answer to that question was how our qualities blend to make us, us. Not everyone is the same amount of kind or arrogant or helpful or courageous. Some are more than others and some not at all. If she had the superpower to look inside someone's soul, she could see how their qualities blended and how exactly they were unique. That's why she wanted to visit the lake. It was the closest she'd ever be able to stare inside someone's soul.

And that's why I had that dream.

What would she find if she could stare inside my soul? I am so broken that I'm not even sure if she'd be able to see it properly. But what would my qualities be? Asocial, dull, dependent, insecure. Those are the words I'll use to describe myself. And those are the words others would use too.

Naisha wanted to go see that lake and now she won't be able to. She won't be able to do anything that she wanted to, she'll never be able to cross something from her huge bucket list or make up new theories about the world. And what was I doing? I was sitting here wallowing because I was too dependent on her and now that I've lost her, I don't know what to do with myself.

Yes, I hate being the way I am. I've admitted that to myself several times. But I've never done anything about it. Why? Because I was always waiting for someone to come and rescue me, be my knight in shining armour just like she was. But I can't do this anymore.

I know the person that I've become, the person I have been for the past several years isn't me. I've covered myself under layers and layers of self-doubt and dependency. I've put up masks on my face, masks that hide who I am. Masks that help me disguise my true self from the world and save me from the task of building up the courage to reveal myself. Masks that drain me of emotions and make me numb.

But now I'm tired. I'm tired of having to be this way. I don't want to be alone and miserable and dependent anymore. I want to be the Kiya that I was, not the one that I have become so comfortable in pretending to be.

I want to change myself. I want to accept myself, with all my flaws and weaknesses. I want to reach true acceptance, of myself and everything around me. I want to remove the mask.

So I take out my notebook and open a fresh page. And on it, I write, asking myself,

"How do we reach acceptance?"

"How do we remove the mask?"

Denial-III

Acceptance

A very simple word with a lot of meaning. If you truly have it in your life, acceptance of self and everything around you; acceptance of the fact that holding on is important but letting go even more. Acceptance of the world and the universe and that you don't control it, you probably have everything that you could ask for. If you don't have it, then chances are you are probably punishing yourself for something you can't control. Trust me, I was there. And in a way I still am.

Acceptance is a long process and it takes time. I'm still trying to reach it, still trying to get past everything that I feel and reach the last, final stage. I know things won't magically become better when I get there but I'm hoping that my journey to the final step gives me enough lessons to be able to deal with whatever comes after.

I'm still trying to accept my body, flaws and all. I'm still trying to make myself believe that I'm beautiful and worthy of love and every other happiness life has to offer.

I'm still trying to regain that confidence I once had. I push myself every day to get through the barriers called self-doubt and put myself out there. I'm trying to make friends, people I can rely on but not become dependent on. I now sit somewhere in the centre of the class and not my usual corner desk of the last row.

I'm still trying to accept the fact that things will not always work out in my favour and when they don't, I must face the problems with my all instead of giving up and pitying myself.

I'm still trying to find my purpose in life, still trying to figure out what exactly I was put on this planet to do. I'm nowhere near the answer but I know I'll find out one day. Until then it'll remain in the part of my mind that stores my great unanswered questions.

The one thing that I have accepted though is the fact that my best friend is dead. It took me a very long time but I've finally accepted it. I've accepted that I won't ever see her again or speak to her again. And as much as it hurts, life goes on. I can't and I shouldn't stop myself from moving on. I need to live my life and not just for her. I need to do it for myself too.

I'm trying to remove my masks, one day at a time. I'm trying to become the person I want to be and I'm trying to let go of the one I used to be. And it is difficult. More difficult than I could have imagined. But that won't stop me this time. This time I will become a better version, someone I look forward to seeing in the mirror.

And how am I removing the mask? By reminding myself every single day that it'll be worth it. All the pain and the vulnerability of becoming authentic will come back as happiness and strength.

Every single person around us is wearing a mask. Some are trying to cover up their pains and others have made pain their mask. Whatever it is that people are trying to cover up, the truth remains that everyone is wearing a mask. Every single one of us. So by removing my own, I want to add one more person to this world who can give others the strength to get rid of their disguises.

We're all damaged and drained in our crazy ways and it is this damage that has made us put on all these masks. We are our monsters, our demons. We are fighting ourselves every day just so that we don't have to see a vulnerable figure in the mirror every morning. These masks that we put on aren't for the world they're for ourselves. And it's about time that we remove them. Because if we can't be real with ourselves, acceptance or no won't matter.

Guilt

– Navya Sheoran

Guilt-I

Guilt, the third stage from the seven stages, a small word that can break a well-functioning healthy human body. It can be paralyzing for some people. Feeling guilty is a default state because deep down I believe that I am bad. It is this destructive level of guilt that I want to examine because constantly feeling guilty is paralysing. A person can feel guilty for something they did, for something they didn't do, for something they thought they did or for not doing enough for a person. When you are guilty it's not your sins you hate but yourself. People say that no amount of guilt changes the past, but the smallest amount can change you.

It's human for every one of feel guilt at least once in their life. I tend to feel more deeply and intently than others and am more aware of the subtleties. I sometimes get overwhelmed by the constant waves of social nuances and others' emotional energies. For me it's like, go bravely; go deeply- or do not go.

I don't understand it myself. I feel different. I feel more. Everything hurts. Everything. I am super sensitive. I feel things that other people probably don't. I hate to see the hate in the world. I want to help everyone. I rise by lifting others. Carrying the weight of other people makes me feel light-hearted. I don't make myself a priority. Never. Because I feed myself by helping others. I like to take their emotional burden and help them through anything they want. Or maybe, that's what I think I like. The storm I am going through does not matter, not to me at least. The storm will go away if I help others clear their way, this is what I feel. Or maybe, that's what I think I feel. If I am able to make others happy, it is worth all the chaos and pain it brings to the table.

I pray I pray to the god every day. "Oh Lord, help me, help me to help others who are in need. Give me the power to move the rocks that cross their path, give me the strength to work for them endlessly, for if I don't I feel I am worth nothing. I feel the guilt to not have helped them. Oh Lord, help me save myself from the guilt."

I remember the day when my so-called "best friend" texted me. Trust me when I say that friendship is everything for a teenager like me. At this stage, when our parents expect us to act like adults and treat us like kids, it is only another person that is my age who is also going through it will understand me. My friends are everything to me, I sometimes even wake up and go to school just because of them. And since they are a huge part of my life, a part that understands me, me who is still confused at who I am and what is it exactly that makes me myself and what do I want from my life. Right now my parents are only a support system to me, but my vital system is the other teenagers who cant feel the home as home, who cant exist for they don't know the purpose of existence, who don't like the society because they don't accept them as the growing mess they are. It's my friends who I look up to when I am in a problem because I am sure they will understand me and look at it from my point of view rather than telling me how small the problem is or how overdramatic I am being. Which is something completely opposite to what my parents will do. They will probably be angry at how inconsiderate I am being and how ungrateful I am for what I have and how I apparently create unnecessary problems. They expect us to understand why they do what they do but show no effort in return for the same. Just because our problems are more related to the feelings does not mean that they are stupid. My parents always say, friends will never stay forever, all this time you waste with them is all useless. It hurts, it hurts a lot. When they want us to open up to them but when we do, we are met by a cold brick wall. I know that they are doing this for my safety because they are worried about me, but they need to know that this does not help. This is not just with me, it's the case all over. Even my friends' parents are not better. They have the same mentality, "friends will never get you anywhere in life". I think this is the reason there are so many "fake friends". Maybe because everyone is taught that one day the other person will leave you so you do not put in any effort.

But not for me, I think that the blood of the covenant is thicker than the water of the womb. This is why all my friends have a special place in my priority list. I make sure to always be there for them. So, as I said, I received a text from my friend at around 8 pm. At my house, you are not supposed to text or call someone later than that because that's, well late. As I read what she wrote, I instantly knew that there was something wrong, what? How? Why? I had no clue but I know that there was something that had happened or something that was bothering her. I immediately called her. But there was no answer. I called her once, twice, thrice but every time there was no answer. Me being the paranoid and the negative thinker I am, I thought of the worst of the possibilities and soon started to panic. Tons of texts were sent, tons of calls were made but there was no sign of her.

On one hand, I am trying to get in touch with her and help my friend, who needs me right now, and on the other, I am hiding from my parents so that they do not see me using my phone. The panic from both collided and turned into something I had never experienced before. I could feel every breath that went through my nose, I could feel every second passing by telling me how precious it was, I could hear the noises coming from the other room that indicated that I can be caught, caught trying to help a friend. My hands were shaking typing the texts, I suddenly forgot how the technology worked. My brain was a complete mess. I could hear my heart throbbing and it's sound so amplified that my head hurt.

I called her again. After a few seconds it stopped ringing, but this time because someone had answered it. I let out a breath which I didn't know I was holding. I said hello. Panic and urgency so clear in my voice that one could possibly see through it. As I waited for a reply, a word at least, I was greeted by a sob. It broke my heart, it hurt. It hurt physically. I could feel the pain in my chest. I had so many questions for which I needed the answers, so many words to say to her but now nothing came out, my mind was completely blank, as blank as a canvas of an artist that is feeling too much simultaneously that he cannot do anything.

After almost a minute of listening to her silent sobs, I exited my terrace and came back to my senses. I asked her what's wrong but never got a reply. The sobs continued and that's was all I needed to want to go to her and be there for her. I wanted to hug her and protect her from everything dangerous in the world. But as much as I did want to do that, I knew it was not possible. Wait, was it really not possible? I could run to her house, it's not that far or I could take a cab, or even better, I could take my scooter. I thought these thoughts to myself. I knew that would be insane and risking everything but I felt it was all worth it.

Suddenly there was a loud bang sound, the sound of reality, the sound of my dad opening the door to my room. Oh, how could have I forgot, I am still a teenage girl under my parents watch. Every nerve in my body went numb. As an instinct, I immediately ended the call. I saw my dad's expression and one thing I knew for sure was that this time, I messed up real bad. I have been warned about not using my phone several times, but then, there I go again. I regretted it at an instant. The tension was thick and awkward. I tried to put on a smile and casually keep my phone on the couch but it was of course too late. Dad had noticed it. I could see on his expressionless face the anger, which anyone else would have missed easily. He said nothing but I do not these

there was any need for that too because him and I both knew what was going to be said. He just took the phone from my hand and went to his room saying am not getting it back.

That was it, I failed. I failed him as well as my friend. I failed to be a good daughter as well as a good friend. I knew she needed me now but I also knew what I was doing was wrong. My dad hates when I call my friends at night. It was as if I could see everything going south but all I could do was stand and watch. It's as if you can see the vase falling but you know you cannot save it even if you run. I stood there motionless, not even my breath could be heard but inside it was so loud. I was so loud that it hurt my ears but somehow no one else could hear it. I was screaming but not for myself. I felt bad but not for myself. It was for my friend who deserves better, better than what I could provide her, someone who could hear her when she said. A friend she could be proud of. It was for my father who deserved a better daughter, a daughter who did not break his trust and understood him. A daughter he could be proud of. I realised I had failed them. Not once but many times. I did something that could not be undone.

I head a shout of my name, it was my mom calling me for dinner. I went to the room, putting on the happiest smile. A smile no one could ever call fake. I smile that I had learned was important for one to have in order to survive in this world. I sat there eating my food as if it was a normal day, which in fact it was. A normal day of showing the world a smile that was there but still missing, sharing the happiness which you do not really have. As usual, I complimented my moms' food, but that day I did not even taste it. I ate more than usual but still, no flavour was tasted. I finished it and swiftly went into my room. Just standing there staring at the blank wall wishing I was like that. But it too couldn't last long as I heard shouts from the other room. It was my parents' shouting at my brother for god knows what reason. Although am pretty sure it does not matter, I still have to go and study.

It was a kind of a routine, a pact, a ritual, whatever you may call it. If one of the siblings is getting scolded the other one goes and studies. I too follow it and opened my book. I wish I had not done that. For then I wished I could be like the book. Serving its purpose and certainly containing knowledge. I tried to read, oh trust me I did. Not for the purpose of learning but so that I could divert myself. Divert my brain and preoccupy myself so that I do not have to hear what's coming my way. Even from the closed doors, even through the thick walls, I can hear every word. Oh, I wish again, I wish I did not have such sharp ears. I wish I didn't have them at all. Then I would not have to listen to anyone, I would not be expected to stand up to their expectations, I would not have to listen to them saying how I failed them. I wish all my wishes would come true.

I sat there, trying to study. For hours or maybe those were just a few minutes. I seriously do not remember. All those shouts from my kept lingering in my thoughts reminding me how useless I am. I clearly remember than stating what an insult I am to them. What a shame it would be to have a child-like me, I thought. Not of any use. How hard it must be for them to show me to the society and admit to me being a part of them. I also remember them saying what a waste of their resources I was. At first, I did not think of it much, but after hearing it almost every day I realised that they indeed were true. All these facilities when I was being provided with, all this love I was given, I was wasting it. I knew they held me guilty for it, and I of course was and still am. Somewhere I knew these were just words formed out of anger and they might not mean it. But aren't the words of anger the words that you mean? Aren't these words your true feeling? People say to listen to a person carefully when they are angry as that is the time they reveal what they have always thought of you. After they had completed making me realise how everything was my fault they ended their shouts. Or maybe just because they were tired of saying so many things and the list would never end. Or maybe because they trusted me, trusted me to be worthless and a waste.

I went to my bed knowing very well that I am getting almost no sleep tonight. I usually do not like to sleep, but that day was an exception. I had never been more eager to feel myself in the dream world, to leave the reality of me and of others which was a complete mess because of me and shut myself down. I understood the meaning of desperate that night. I understood when I felt that. The desperate need to stop it, stop everything. To detach from the reality so that I can no more disappoint or hurt others. I knew sleep was playing with me. I was tired, physically and mentally, of myself. So tired that I could not even wipe the tears that were falling of my eyes. Although I was sure I had to sleep immediately otherwise I would do something to help others, help them to get rid of me. The nights like these time holds still as if waiting for me to do something. The sky becomes rather bright than usual and stars come out as if an audience waiting for the show. They watch me as I drown in my own guilt, sitting there far away. I close the curtains as I do not want anyone to see me cry, not even the lonely stars.

I cover my mouth so that I cannot be heard by my parents and not even by helpless the street dogs. I know it was my mistake and I have no right to cry. I know this is what my parents will say. Who are they to blame, even I do not what to hear my ugly cries after hurting others. The clock continued ticking, the stars kept shining and I kept thinking and thinking and thinking. After a while, the stillness stopped bothering me. And then came another wish, which scared me somehow. I wished to become still. I knew I had no right to fulfil that wish because I deserved what I was going through. It will not be right to leave everyone in the pit I threw them into and escape alone. It would make me feel GUILTY.

As I was busy thinking of my escape as any selfish person would do, a thought so casually came to my mind. The thought of her. I imagined her in my place and myself in hers. Wait that does not matter. It was the same. This is what I was trying to save her from. This is what she was going through when she texted me. The train or my thoughts sped up. So fast, it was going too fast.

She was here, she felt this, she was blaming herself, she was watching the stars just like me, she was listening to the ticking sound just like me. She was blaming herself just like me. She wanted to end it, just like me.

Guilty. That was what I was. That was what I felt. That's what she must have felt. I left her there, to cry herself to sleep. I left her there in the death trap. I was guilty of letting the guilt consume her. What have I done? I could not help her. I did not do what I was supposed to do. I disappointed her. And not just her I disappointed my parents too. Everything was my fault. Wait, this is where it started. It is starting over again. No, I do not want to feel it again. I do not want to feel numb. I do not want to feel nothing. I knew I wanted to escape, for this will go on forever. It is never-ending. I told myself I need to leave, I had to leave. But then there came a voice, it was not so sweet but it said what was the truth. I had to suffer for what I did. The voice, it came from inside my head.

I do not remember falling asleep that night but I certainly did wake up the next morning. The alarm seemed not to harsh that day. Oh, it was not the alarm, it was my mom who works me up. Even through my half-opened eyes, I could see her happy morning face as she handed me the mug of milk. From the other room, I heard my father laughing heartily with my brother. He came to my room and greeted me. Asking if I want to join him and my brother for a swimming session. Hope and happiness clear in his eyes. I could not say a no.

I quickly finished my milk and ran to the washroom. It was the morning and I had to put on the mask again.

Not surprising to me anymore that my parents did not need one. They had already got over what happened last night. It was of course not a huge deal. Who would say it was? I should not be crying over what happened yesterday. I was not even scolded directly by my parents. And it was such a small deal anyway, why would I even think about it ever again.

Guilt-II

And with that thought, I went to the swimming class. I knew swimming would give me an escape, even if for an hour, but it would. As we reached there, I noticed how few people were there, fewer than usual and that made me happy. I got excited and ran to the pool. The clear water felt so welcoming, the sun rays reflecting from it making it look so beautiful. This happened long back but I clearly remember that the pool didn't look like a pool that day. Even with the same environment, it looked much more. The water inside moved rapidly and there were almost waves caused by the people inside.

Ocean, the pool looked like the ocean that day. It was smooth yet powerful, with enough force to destroy and rival the land. I didn't understand if I should revere it or fear it. I slowly descended into the cold water and started swimming. You know, I liked it here, I didn't have to bear my weight, the water would do it for me. I didn't have to try and stay on the surface, the water would keep me there. I trusted the water too much, I can't help it. I just loved it too much. It embraced me completely, too well for my liking. I wanted to stay there. It's as if the water wanted the same. Maybe I could do that, stay here forever, and be with it forever.

At that moment, it seemed so right. There wasn't even a single doubt or second thought. How it would feel if I let the water into my lungs, how it would feel to go limp, how it would feel to feel nothing. It made me curious, it made me desperate, I wanted to do it. I wanted to feel it. And I did what I wanted. I went to the deeper side of the pool. "Perfect", I thought to myself. And thinking that was my last thought, I went into the water. I went under deep, I went till I touched the ground. I was about to breathe in the water, but something stopped me. It was just me being a coward. "should I do it? should I not? This is bad! But this is what I want right?"

All these going through my head non-stop. But I came out, up to the surface. Coughing and crying. Out of breath, out of water. "I couldn't even do that. What a coward I am."

As I was cursing myself, I felt someone pulling me up, out of the pool. I little bit later when my vision cleared I recognised that it was my dad. I could see the worry in his eyes. The emotion dripping from his face. I felt bad. Bad for making him feel that. But it felt nice. That there was someone who actually worried for me. "Wait... What am I thinking?" How could I be so selfish? I was enjoying the feeling he got. Such a manic I am. He examined me and asked me what happened. "Oh, nothing! I just slipped." I replied and quickly went to change. I came out in my school uniform. Yes, I had school after that. Something which I wasn't ready for at all. I was not ready to face my friend. Not after what I did to her yesterday. I felt ashamed, I didn't want to meet her nor even think about her. But then I can't be selfish again. I had to check if she was fine. I had to see if she even showed up to school or not. I was worried about her.

After my dad had dropped my brother to school. It was my turn. My stomach was turning upside down. I could throw up any second. I felt it again. As if I was in the water. I couldn't breathe, except, I had air this time. I thought of telling my dad to take me home because I didn't want to face it. I was scared. But by that time I had already arrived at the school gate.

"Bye kid. Take care." My dad said before driving off. I stepped into the school. There were so many kids, some happy to be here, some not. Some walking in groups some alone. I didn't want to pay attention to any of it. But I did. As I already told you. I observe a lot, and sometimes it's not a blessing. I walked to my class. Each step as if took me an hour. But that was too fast for me. Each stair I climbed, took so much energy, but that was still less for me. I climbed up to my floor. As soon as I was there I was rammed in between bodies. What were my friends, and this was apparently how we greet every morning. I hugged them back. It felt nice. "keep your bag in your class and come out quickly, we have so much to tell you!" and with that, I was pushed into my class. And I did what I was told. I kept the bag on my seat and happily went outside. I was smiling from ear to ear. I went to my two friends. And without wasting even a single second, they started about how their day went yesterday. They kept on talking and talking. Their words never-ending, their mouths never getting tired, their hands making all kinds of actions. It's as if they had chugged down a whole ton of some energy drink. I seriously need to know who supplies them with all this stock of the drink.

But I loved to see them like this. All happy. I loved to listen to them every day. They had so much to say as if we met after years. Research says that people are most scared of public speaking after death. I don't think that applies to my friends. They could talk all day long, to anyone and literally anywhere. And I dare not give them a mic and a stage. Imagine the state of the poor audience.

But their current audience loves and supports them so they are alive. And I gave them the same enthusiasm back that they were talking with. I listened to each word of theirs and gave my view on every topic being discussed, but they apparently didn't have my full attention. I kept taking glances to the other class. Where my other friend should have been by now. "She came to school right? Nothing bad happened right?" But I didn't let it on my face. I knew I couldn't worry these two with what was happening with me. The chatter was stopped when the bell rang. We were supposed to be back in our class. All the busses had arrived. This could only mean one thing, she didn't come to school today. My fears were coming true. I felt sick in my stomach as I went to my seat. I was worried, not for myself but for Siya. She is very sensitive yet so brave. Yesterday night she would have done things which I was scared to do. The feeling that she might have done something to herself didn't leave me.

I was so determined and concentrated in panicking that I didn't notice my teacher entering the class. It was when I saw her that I realised it was my economics lesson. One of my favourite subject. She started with her teaching and I started, or at least I can say I tried to listen to her. Although the class was quiet except her, although there was no other sound, although she was speaking loudly, I could not even listen to a single word. I tried, I tried so hard to concentrate. I knew I lost track and was trying to understand what she was saying but I could not and that just made me angry. All she said fell on deaf ears. I could hear her but I wasn't listening to her. I looked at the board and noticed that we were supposed to do a numerical. Having no idea how to solve it I just copied the question and sat there staring at it. Never had a blank paper amazed me so much. After a while, I saw my teacher walking towards me. "Dammit, I am doomed!" I thought to myself. I quickly picked up the pen and acted as if I was thinking hard.

My teacher came and stood beside me for a minute or so after which she said,

"What's up with you today girl? Why can't you solve the numerical, this one is quite easy?"

I wish she would have asked what had happened to me rather than why I was not able to do it.

"Oh, nothing ma'am! I just got a little confused with a step."

"I am no fool darling, I can see something is bothering you. Are you okay?" she asked with a concerned look.

That's when I realised that this was a bad idea. I didn't want her to ask me what happened to me, for after hearing this I might break apart, for I might lose what I was holding in.

"I am completely fine ma'am. Thank you for asking." I quickly replied in the hope that she would leave me alone.

"Ok child, I get it.

(No, you do not. I thought to myself)

But feel free to talk to me if you need any help."

With that, she left me there with my blank sheet. The whole lesson I caught her taking glances at me. Any time we made eye contact I would pass her a happy smile. This carried on till the bell rang indicating the short break.

I ran towards the classroom door wanting to get out as soon as possible. My classmates looked at me confused as I bumped into some of them. It was really a sight to watch. It looked as if I was being chased by a dog. I wish I was funny to me, at least I would laugh, a real laugh. I opened the door and ran to the other class.

The students there were still inside as their teacher hadn't left them. I looked through the small glass of the door. One by one I scanned the whole class. Student after student, face after face I searched for the familiar one close to my heart. Heart, my heart was beating so fast as if it would tear off my chest and come out. Well, not my heart but the teacher did come out. As she took her steps out of the class I rushed in.

And there she was, sitting in the corner of the class staring outside the window at the beautiful scenery. It was so beautiful outside, the lush green trees, the singing birds, the bright sun, exactly opposite to me as if nature knew and was teasing me for the happiness I didn't have.

I slowly walked to her, not wanting to scare her away. I tapped her shoulder so lightly as if she was made of glass, so delicate that she would break if not handled carefully, so delicate that she would break by a mere touch. Without even looking back, she threw my hand away harshly. With that one action, she broke me. Wasn't I the strong one? How could this be possible? How could such a delicate creature do that? I was confused, scared, angry, sad…..and GUILTY.

I could feel the hate radiating from her, hate that was just for me. "Siya….look I am so sorry about yesterday I can" before I could complete the sentence she got up from her seat and gave me a sharp, stern look. I didn't exactly know but I guessed that she must have felt so bad, so disappointed and much be so angry at me. Without a single word being said she left the classroom. It took me a few seconds to realise what had happened. And as soon as I did I ran after her. I followed her to the washroom which was strangely empty as it was the break and mostly there are tons of girls here. Dammit, I noticed again. These small things I notice make me feel vulnerable and naked to the world.

"I know you are mad but please let me explain!" I cried to her. But she didn't even spare me a glance.

"Hey don't be like that! Siya!" Still no response. "I know I did wrong! But please don't ignore me. Please talk to me. I will tell you everything. I have a valid reason." I tried to talk to her. I tried to make her listen. But my words were falling on deaf ears. Never had I wanted a response so bad. At that moment I felt helpless, I could not do nothing which would make her hate me less. I wanted to tell her I feel sorry for what I did. I wanted to apologise but she would not listen to me. I felt as if the only way to escape was by cutting through it but my hands were tied. I tried so hard to break the wall she had built around her. I begged, I pleaded but she didn't move an inch. She was standing there staring at me not even getting affected a little bit. There wasn't even the slightest emotion or sigh that she will forgive me. Not now, not soon, never.

After almost ten minutes of me literally begging to her, she talked. It wasn't what I wanted, although I expected that, it's not what I wanted to hear, not now. As soon as she opened her mouth, her words made it clear that I was a disappointment to her, I had messed up again.

"REALLY?! You want to apologise? Do you think that would help? I was so messed up yesterday! I needed you and you weren't there. I was counting on you to look after me. Look after a friend. This won't help. This apology means nothing. You don't even consider me your friend otherwise you would have been there for me."

I was listening to it. Listening to every word she said. I wanted to know what I did wrong. I wanted to help her.

"I know what lame excuse you will make! I didn't have my phone or I was eating or that your dad took your phone. Can't you even ask your parents not to do that? Not even for me? I am supposed to be your best friend. Is this how you treat me? I knew I should have never trusted you. I am always there for you when you need me. Have you ever seen me say no to you? Tell me! I don't even know if that is true or you are just making excuses so you don't have to talk to me."

I could not believe that she was saying that to me! How could she? "You cannot say that. I was seriously trying. I was super worried about you." Again before I could finish I was cut by her words, or should I say sword. Well does not matter, it felt the same anyway.

"Don't even say that you were worried. You have no clue what I went through last night. I felt so sad. I wanted help. My parents wouldn't listen to me. My other friends were not available. I thought I could trust you, I thought you had my back. But that was certainly not true." She said with rage in her voice but hurt in her eyes.

I heard her voice crack in the second sentence, and after that, I stopped listening. I didn't need to. I could see what she wanted to say. No word could explain perfectly than what her expression did. No hurt of her words could compare to the hurt of her voice cracking. I saw how her tears filled her eyes. That was the last sword. The last sword that pierced right through my heart. I felt it physically being stabbed into my chest and being twisted with a huge force.

While I was recovering from the stab, she pushed past me and ran out of the washroom with her last words

"Do not talk to me ever again, you mean nothing to me."

I stood there in the washroom feeling the walls crushing me. The amount of emotional pressure I felt at that moment was so unbearable. The school uniform started to suffocate me. My reflection in the mirror started to disgust me. I could not look at the ugly, unfaithful creature standing there. I mean look at me, I still had the audacity to go and talk to her after what I did to her. Did I have no shame? I turn on the tap and splashed some water on my face trying to erase the memory, trying to erase me. I looked at myself again; I knew I was at the verge of an ugly cry. I can't do that now, well at least not in school and also not in front of my family at home. These cries were only heard my pillow at night, the tears were only revealed to it for it did not ask.

My pillow never asks me why I am crying, it never asked me why I am so weak or what did I do wrong or why did I break before helping others. It never says that my problems are useless, never have I once heard it refuse to listen to any of my cries. It just absorbs all the tears till the last drop. It absorbs all the sorrow and the worries in my life. I wish I could bring it everywhere. It lets me be myself. It makes me feel like it's okay to have a bad day, it's okay to make mistakes, it's okay to be less perfect, it's okay to do what's best for you, it's okay to stop worrying about things I can't control, it's okay to be yourself. I love my pillow. It helps me through my life. But sadly I only remember it in the times of need. Sadly I only go to him crying. Sadly I never worried about what it was going through. Sadly I never care what he feels. Sadly I do not care at all about what happens with it because I know that it will be there with me at the end of the day ready to listen to my cries quietly.

How bad must the pillow feel to be treated like this? Trust me I know. The pillow does everything it can do for me but still feels it could do more. The pillow does not hate me for coming to it every time with tears, but it hates me for not wiping his tears. The pillow never feels burdened by my words, but it feels burdened by itself.

I don't know how long it will hold together because it has absorbed a lot till now. I don't know how long it will last before it bursts with the excess emotions. I don't know how long I will last till I burst. I don't know how much longer I could be a pillow to others. I DO NOT KNOW HOW LONG COULD I LIVE WITH THESE PEOPLE.

My internal screams were hushed by the opening of the washroom door. I saw a few girls enter and I knew I had to leave immediately. I looked at myself one last time before leaving the washroom. As I went to the corridor I saw no one there. The next class had started and I somehow didn't hear the bell ring. I ran to my class as fast as I could. I reached and opened the door.

I saw the teacher inside. Psychology, it was psychology lesson. I looked at the class, everyone was staring at me as if I had killed someone and had blood on my face.

"Come on! I am like a few seconds late" I thought to myself and looked at the wall clock. Well, it could have been a couple of minutes. Okay, I was almost 20 minutes late.

I quietly went and sat on my seat. Although the burden was still felt, I also felt a little hope that moment. How amazing it is how fast the emotions change. I am quite close to my psychology teacher. Not because I have been her patient or anything but because I just like her. She is always there for people and ready to help. She is an idol to me, although she has no clue about it. I want to be like her, strong and light in people's lives. We talk a lot, about everything. And that's why she let me in without saying anything in case you were wondering. But I was never her patient and never asked for her help because I didn't need it. I am a very positive and optimistic person. Well, that's what anyone who "knows" me would say. And I would happily agree.

But today was a bit different.

At first, I did not want to admit it, maybe it hurt my non-existent pride or maybe I was just too foolish or maybe I thought I didn't deserve it. Even today I will never agree that I felt all of them be true. It was true that I was afraid of help. As believed in psychological theories and perspectives, 'Known pain is better than unknown pleasure.' Well seems like I too was scared of the pleasure, of help because obviously I had never received it before and was too scared to receive it anytime soon. I was not worried about what would the people think, I was worried that it would prove I cannot handle myself or that I am not a happy person which I strongly believe I am.

But as I said that day was different. I somewhere inside my heart wanted her to somehow notice me and help me. I was just sitting there in the class with my open book staring at my teacher hoping to get a single glance from her. I fiddled with my pen, flipped the pages of the book, dropped my bottle, and coughed a little; all these small things to get her to look at me. I did everything I could do to get her attention. But today being different she didn't notice me at all. Somehow she didn't notice my call for help. Was I not clear enough? I think I was. With every passing second also passed my hope. As every second slipped so did the chance of me getting help slipped away. I was literally staring at her after a point of time. And almost after five minutes of that, she looked at me.

"Yes! This is it!" I thought to myself. It was not a long eye contact but enough for me to send my message for help. But guess what I did? I smiled at her with my eyes so bright that it could have possibly blinded her. What a foolish and coward person I was. Can't even ask for help. I tried to calm myself down and blame it on the noisy class which was sitting quietly and on my teacher who was not ready to help me. Because this is what we do. This is what we are taught to do, blame others for our weaknesses and the condition we are in.

But even after this, I didn't lose hope. I had faith in her, in her job. She is a school psychologist, she is supposed to know. She is the one who should be able to listen to the silent cries of us, of me at least. Even now some part of me did not want to go to her asking for help because she already has so much work, she has too many children coming to her every day with her problems and she also had a personal life and a family to take care of. I am supposed to be her help and lighten up her mood and not the other way round. I could not add up my burden to her. But I also wanted her to help me, not too much, but just little enough to make me get back to my normal self. I wanted her to notice me crying. I thought to myself again.

"She is a psychologist, she should see more than what meets the eye, hear more than what's being said, and feel more than what's being put forward. She should be able to see through the mask which people have been hanging on their faces. A mask that is soundproof to hide their cries from the world. She should know, she should help. Help people who are afraid to ask for it and know that they need it. People like me. She should help me. Please help me. I do not trust myself anymore."

I wish she would but as I know, none of these wishes come true. And this one also piled up in the list of wishes. Only 10 minutes were left. "DAMMIT! I need to do something!" I whispered to myself. Is the clock working properly? Was time always this fast? My thoughts were faster than the fast ticking clock. Everything was a mess. There were too many thoughts. Their speed was so fast that it could beat the speed of light.

While I spent my time controlling the over speeding thoughts, the lesson was over. I saw my ma'am taking her books and leaving the class. Wow! It looked as if everything was happening in slow motion. It was too slow for my liking. It was as if nature wanted me to go to her as if it had slowed down just for me and tell me to do what I was afraid to do. But I didn't. I just stood there and watched her leave.

"There goes my hope." I thought of running behind her. I thought of going and talking to her and tell her my need for her but I didn't. I was afraid she would not notice just like she could not a few moments back. And it was too late anyway.

I sat back into my seat waiting for the horrible day to go on. At that moment I just wanted my pillow. But not all wants are fulfilled. Soon I heard a shout of my name from the corner of the class. It was my friend, Yuvraj.

"Come on take your file! What are you waiting for? It's the art lesson." He said with the happiest voice I had heard all day. I looked at him and returned him the same happiness. But at that moment I was burning on the inside with the feeling of jealousy. How could he be so happy? I am not I person to get jealous easily but this time I was over the top with that emotion. I wanted him to lose his happiness just like I lost mine; I wanted him to suffer just like me. He also deserves the pain. If not nature, then I will do it. I will take it away and break him. I WILL TEAR HIM OF-

Wait, what am I thinking? What has happened to me? I am scaring myself.

I just took my art file and started walking with him to the art room. I tried my best to hold back my new upcoming ideas which I did not like at all. I am glad he didn't notice how it took me each one of my 100 nerves in my body to control myself from saying or doing something I would regret. He kept on talking about something or the other as we exited the home building and entered the junior wing where our art room was. I didn't mind him and brought my mind back to what had happened earlier, trying to find out its cause.

"My psychology teacher is very experienced and one of the top in her field, there is no way she would not have noticed me. I expressed so much. I ut I did whisper. Maybe she didn't want to help me. Maybe she knew I was not worth the time and energy." Yes, this must be it. She didn't want to help me because I do not deserve it after what I did to Siya. As these theories were forming I thought I wasn't listening to Yuvraj until something he said hit me. It made me stop dead in my tracks.

"Art is my therapy. It simply makes me feel better. It's my emotion; it speaks the works that I am unable to explain."

Even he was wearing a mask. And I got fooled by it. He also is a pillow to someone. He also feels what I am feeling. But just like me, he is hiding it behind a mask. We both are hiding the same thing from each other and waiting for each other to notice it. There is not much of a difference between the two of us. It's just that, he still has absorbing power left whereas I am on the verge of bursting. He still has his mask tight on my face but mine has started to slip off because of the tears. This is one of the biggest risks. These masks only get off by the salty tears, so every single tear that falls from the eye starts to slowly slip off the mask. And trust me when I say it is very difficult to get the mask back on once it is removed.

I reached the art room with my mind overflowing with guilt. I was guilty of not understanding him earlier. I wish I could have noticed his call for help and been a pillow for him. I wish he didn't have to suffer like this. But it was of no use now.

I quietly walked to the table where I was working because I didn't want to bother anyone anymore with my little guilt phase. I opened the bright coloured paint bottles but I somehow didn't like them anymore. Their brightness, which at one time amused me, is now not that soothing anymore. I dipped the brush in the paint ready to complete the half made galaxy. I leaned back I little to observe before starting. Oh, how much I wish I hadn't done that. The view of the incomplete yet so beautify galaxy blocked my mind and I liked it. Looks like art is also my therapy and I was foolish enough not to notice it till now. I had made the sky that supported me. It was the same sky I looked at every night from my window while I cried to my pillow. Those were the same lonely stars that I wanted to hide from. Then I wished again. I wished I was could lay in the sea and look at the night sky for the rest of my life. Far away from everyone. Just me and the millions of lonely stars. Because now I wanted to show to them that I was their part too. It was as if I was hypnotised by the galaxy. It was so beautiful I wanted it to consume me; I wanted it to destroy me. I felt the burning tears building up in my eyes because of the beautiful sight and the thought of the galaxy destroying me. It is a fact that we cry because of the excess of emotion. There is just so much of that emotion that our body cannot handle it and somehow has to take it out so it uses the method of crying. And I was crying because I felt beauty. I felt the delicacy of nature.

At that moment I was remembered, Van Gogh. What an amazing madman person he was. Even if he tried to eat yellow paint because yellow is a bright colour and he thought eating it would make him happy. Although everyone thought he was crazy for eating something so poisonous and toxic. But if you look at with less judgemental eyes you will see that he was just trying to stay happy. He was ready to poison himself for the little amount of happiness. This made so much sense to me at that moment.

Although I could not eat paint, I had other ideas in mind. I knew I could save me. I didn't need no one else. I will be my own pillow. I knew I had to take things in my hand. I will do what Van Gogh could not do but in my own style. I went to the back of the art room where most of the supplies are. I easily found the metal scale and snuck it under my skirt. I was so excited! So happy! After that, I went to my teacher and asked her if I could use the washroom. She allowed me and with the biggest and brightest smile. The smile, it wasn't forced. Now I knew. It was a real smile. I probably looked like a creepy maniac at that time but I didn't care. Rushing through the corridor I reached to the restroom. I quickly went inside a stall and took out the scale. I carefully touched the sharp end of the scale. It left a little mark on my finger. I liked it. It was perfect.

I slowly unbuttoned my shirt and let it hang on my shoulders. My skin looked a little pale. Maybe because it has sensed what was coming it's way. I roughly took out the scale from under my skirt. My confidence was reaching another high which I had never experienced before.

I took the scale and pressed it on my stomach. YES! THIS IS IT!

Then I ran it along my skin tightly pressed. After almost half a centimetre it dug into my skin. I saw it make marks; I saw it make me bleed. There was blood dripping from the corner I started with and carried all the way with the mark. Call me mad but it looked beautiful, so beautiful that I wanted to cry. The sizzling and burning pain caused felt nice. It felt good to be brave. Although I remember having to stop and take a few breaths. It was not because I was scared, but because I was too excited. But I knew I could not stop now. If I did, I will lose confidence. I made another one right below it. This one was a little longer. I made another one on the right side, short but deep. I clearly remember the long scratch marks with blood. It looked like an ornament on my skin. The red beads falling down along my waist. But that wasn't enough. I wanted more of them; I wanted to look more beautiful. I made a small one on my shoulder and one on my thigh. I stood there feeling the blood drip. I pinched the mark on my thigh to extract more blood. The red looked so pretty on my skin. For the first time, I felt beautiful.

My amazing moment was interrupted by the housekeeping as she banged on my door and told me to get out soon. I knew I couldn't go out like this, I wanted to enjoy the show but I knew I had to leave. I wanted to make one on my arm so that I could show it off to the people but I had to stop.

I opened the stall door halfway and looked outside checking to see if someone was there. That day luck was on my side and not a single soul was to be seen in the washroom. Well maybe there were, but I am pretty sure they got jealous of my courage and ran away. I went outside the stall and locked the bathroom door to get some privacy. I took a bunch of tissue paper and wrapped it around my hand and put it under the tap. I looked at the marks one last time and smiled a farewell. I wiped the blood off with the wet tissue paper. I was very careful with everything because I knew I couldn't leave even a single drop as it would stain my white shirt and people would get to know about the little fun moment I had and will try and take it away from me. I would never let them do that to me, ever. I cleaned myself and the scale the best I could and disposed of the tissues. I buttoned up my shirt and make it look normal. I took the scale and put it back inside my skirt. I went back to the art room with a smile. As I opened the door and entered I was greeted by my teacher who asked me what took me so long but I just brushed it away with a small awkward laugh. I knew even she didn't care much anyway. I went to the supply corner and kept the clean scale back. I walked past Yuvraj to get to my seat and whispered a little thank you to him. I saw the confused expression on his face and couldn't help but laugh a little.

I sat back in front of my canvas. The paints looked bright again and so did the outside world. I completed that canvas that day by which I surprised my teacher and myself. It is still one of the best pieces in my collection until now. And so has that day become one of the best days of my life. My beautiful and happy life.

Since that day it became a game for me. I could not play a lot but I did play. I knew that I didn't need the pillow anymore because now I have the scale, my yellow paint. Which apparently was more helpful. Although this does not mean I did not use the pillow. Some odd days I talked to the pillow as well. The scale is a better way but now I have to be more careful. I cannot wear clothes that show much skin but I don't mind that. These are the small sacrifices everyone has to make. I do not recommend this way to anyone, not because I am selfish and want to feel all the pleasure alone but I know it is an addictive path. Even after choosing this path of not using the pillow anymore, I am still a pillow to others. I don't think I'll ever be able to change that. I want to save the rest of the world from using the scale so I have to be the pillow. The scale was the yellow paint to me. Everyone has a different yellow paint. I wish they don't drink the yellow paint. This was my last wish. Trust me it was. Because there is no going back after you take the first sip. This is how guilt made me feel. This is what it did to me. I am still the happy, positive, free, optimistic girl everyone "knows". I haven't changed much if you see me, but I have changed much if I see me.

It was just one of the many encounters I have had with guilt. It's not the best thing that happens to a person. But it sure is a beautiful thing. It makes you a stronger person, physically and mentally. It makes you open your eyes and look from other people's point of view. It makes you feel nothing and everything at the same time. It does things to your body that you have never experienced before. It does great help to you. But beauty has a price. As though it helps you, it breaks you and crumbles you and shreds you into millions of pieces that you know will never join back again. It marks you. Marks you as a victim of itself. Guilt makes you consume your yellow paint till the last drop.

Even if you ask me now, after all this time and after all what happened, I am still scared to face it. Still a little scared of me, for I might not know what I do when I am bored and need new yellow paint. But this time I want to prevent it. I don't want to wear a fake smile anymore, I want the courage to ask for help. I do not want to mask my cries. I want to be able to be brave, not so that I can drink the paint but because I want to be able to reach acceptance. I want to let the tears flow to loosen the mask that prevents me from showing myself to the world. Although I am worried about what the world would think of me after it is removed. Will they hate me for being fake all this time? Will they judge me for all the fake smiles I have shown? Or will they understand me? I have no clue. Although this is it! I am done! I want to remove the mask. But how… HOW DO WE REMOVE THE MASK?

Guilt-III

Acceptance

All this while, I tried to change what was happening. All this while I tried to get people to think that I was strong and then think I was not. I am confused myself. Once time I want everyone to think I am brave and have the courage to face whatever that comes my way and also help others. I want them to know that I never get tired, at least not by helping them. I want them to realise that I can manage life and my emotions and other emotions and their life and my future and……. Everything. I hope they realise that I have the brightest smile that I use to light up others way and that makes me happy. Also, I am happy all the time. Things like stress, anxiety, social expectations do not affect me at all and I do not care about the mask as I don't wear one.

Other times I want them to notice that I do. I do wear a mask and I want them to make me take it off because I can't do it myself. I won't do it myself because I am not that strong. I want them to realise that sometimes I get scared of what comes my way, sometimes I do not know how to deal with problems. Sometimes I can't manage my life. I do get affected by the work stress and

I get anxious very easily. Also, I sometimes get so weak that I stop functioning. And to top it all it makes me feel useless. It makes me feel bad and weak. I feel guilty about that. But I want people to take my mask away and show me who I really am and to accept me as I am.

I talk about change, but as rightly said by the psychologist Carl Jung "We cannot change anything unless we accept it."

If I want to change, that means I want people to accept me and if I want them to accept me, I first need to accept myself. Self-Acceptance is, loving ourselves enough to accept the painful truths about ourselves. It's just being nice to yourself and letting yourself breathe a little. It's just forgiving you, not just once but again and again and again. As many times it takes for you to reach peace.

I learnt it the hard way. I was stubborn and always kept myself on my toes. I always gave other people the top position on my priority list. What I did wasn't bad. It's a blessing to be able to help others. But to help yourself...... it's the best thing one could do for themselves.

And if I asked you to name all the things you love, how long would it take for you to name yourself? You owe yourself the love you freely give away to others. I realised this.

Now I know. It does not take much to love yourself. It's a hard path, harder than that of drinking the paint. But this in the least I could do for myself. Slowly and steadily, step by step, I walked this path. Each milestone I hit, I felt better, better about myself. Soon my paintings were not just about that I wanted to hide from. They became something I never expected them to be, they became a part of me. Every day I paint a portion of myself onto the canvas. Every day I learn more about myself.

I might have lost a lot of people on this track, but I did gain more precious ones. They love me for who I am and more importantly, they make me love myself. If one opens their eyes they can see the sense in Van Gogh drinking the paint. But if one opens their heart they can see the pride in accepting themselves.

I was so desperate to remove my mask that I chose the wrong way. But in the end, I realised, that there is no other way to remove the mask than to accept. Accept what happened, accept what will happen and accept what you have been through all of it.

To come to the end of the process of finding yourself means to return to the starting point. What must be discovered in the end is the beginning of everything, the milestone.

Now I have reached the end. I have reached the starting point. I have reached acceptance.

Fear

– Samira Bhayana

Fear-I

The tear stained keyboard.

3:18 AM, here we go again.

The room was dark, air was decayed, eyes were glassy and my heart was broken. A striking pain was shooting in my abdominal like lightning on an empty road. Fear. Fear had enveloped the room like a whirlwind on a gloomy day.

A whirlwind: "a weather phenomenon that causes terror and havoc to its surroundings." At this moment, from the innocent perspective of a young girl at the tender age of nine, her world was a whirlwind of emotions.

Since I was very young, I had been brought up by the most dedicated, selfless and unflagging parents I have ever seen, along with the most suitable role models, my two sisters of unblemished qualities and hearts of pure gold. Since the very beginning, I had conditioned myself to have the same morals as them, the same strengths, good habits and mannerisms. However one thing I never picked up from either of them was their eternal strength in all aspects of life.

I was the one who was always excused, the one who was pampered and, the one who's tears would always be wiped right away. So, growing up, I never really learnt the art of strength because I had always been conditioned to derive my strength from the ones around me.

The strength however was not as eternal as I thought it was, for the ones around me.

Speaking of eternities, people don't physically stay forever and that was a concept that a nine-year old found exceptionally difficult to grasp and master.

Sundays are the laziest fraction of the week; It is the day where late nights result in late mornings. The day where lunchtime is breakfast time and where all work is put aside only to breathe in the tranquility of a non-hectic day. However, one specific Sunday morning changed the preconceived notion and opinion I held about Sundays.

I was alarmingly woken up by the sound of what I thought was an extreme case of a laugh attack, but turned out to be so much worse. The supposed laugh attack continued and that's precisely when I realized that, that was a case of very extreme case of abnormally hysterical crying. I opened my groggy eyes and to my horror, I see my mother laying on the floor, screaming crying into her arms, wiping her tears on the sleeve of her blue and red flannel shirt, surrounded by the rest of my family consoling her and holding back their own tears.

Though I was young, I knew however, that tears of that magnitude shed by my mother, the single toughest woman I know, meant that something was definitely very wrong.

I looked up over at my father who was right there beside her, also groggy and tossed my head up slightly with a perplexed look on my face, asking him what was wrong.

He looked up at me with a sad and somber face and pronounced the word "nana" with no sound. I understood what he said, but I was in absolute disbelief that that could have even happened that I asked again with that very same curious and perplexed look. He answered the same and I couldn't help but be teary-eyed as it was too difficult an emotion for me to conceal. And so, I let the tears flow and oh yes they did.

My mother being the fighter that she was, got up, wiped her tears, packed her bags, booked a flight and was on the next direct flight to Baroda, without a speck of weakness (she never showed it, yet I knew it was always there, evidently present in her eyes).

She left the house like a fighter, but I knew all along she was breaking on the inside.

When she landed finally, she was given the worst news ever at the worst possible time ever. The doctors let her know that my Nana was no more and that news she did not take well. What followed the news of his passing was a funeral, of course, which was attended by all the people who loved him and remembered him for the righteous man he was. The tears of sadness soon were molded into tears of happiness when we heard all the stories and compliments that his loved ones had to share about him. It truly made us realize that he spent his life here to it's utmost capacity and did all kinds of good deeds for people, and left his mark in the hearts of all.

This is where my selfishness came into play; something I will hold against myself for as long as I live.

I am absolutely and completely excessively obsessed with my birthday, in the least narcissistic way possible.

As a nine-year old, one of the few things you look forward to, is finally entering the age of the widely crazed about "double-digit number". Whether it's the (insignificant) sense of maturity one derives from being a double-digit aged child or just the feeling of pride to be a part of an age bracket that makes one feel more adult-ish and responsible, double-digit ages meant access to the official mini adolescent life.

Now, my birthday being five days after Nana had passed, I insisted on having my 10th birthday celebrated at all costs, even though it was during the very difficult time that my family was going through.

So, I did what every 9-year old would do: I threw a tantrum. That's where my selfishness comes into action. It never hit me that what I was asking for. I was not just asking for a birthday party, I was looking for an excuse to make things about myself in a time that is about literally anyone except me.

What I know now, how I wish I knew then.

I showed no empathy and was not the ideal daughter, and for that mom and dad, if you're reading this: I truly am sorry.

I still remember the day it finally hit me that my nana was no more. It was the first day where the shock of it all actually sunk in.

I was at my best friend's grandparent's house for a Diwali celebration. My parents were away in Baroda taking care of my still unwell Nani and so I was under the care of my best friend's parents for that time being.

While at her grandparent's house, I wandered away from the vibrancy of the crowd and made my way to (what looked like it was) her family room to meet with my more solemn side and locked my eyes onto a wall overflowing with pictures of her with her grandparents. I browsed around the room, glancing at every photo, analyzing it's every aspect, pixel and blemish. That was the exact moment that it hit me: I will never get a single photo with my nana ever again. My eyes felt tinted with a glassy finish over my hazel eye. Next thing I knew, I was surrounded by a large group of people consoling me, stroking my braided hair and encouraging me to not give up the fight of accepting his passing. Once the tears had started flowing, they did not stop till my heat had let out all of its bottled up fear and feelings of melancholic silence.

"Draw a picture of one thing you want most in the world" my grandfather asked me on a rainy idle Sunday afternoon. I, as an eight-year old euphoric and exuberant girl who wanted the WORLD, stared out into space thinking of any ONE thing she wanted most, from the colossal list of wants she had at that frivolous age.

After hours of internal debating, I decided to pick up the red Unibol pen and draw a doodle of a red Scooty behind a fruit bill from our neighborhood fruit seller. Messy, smudged and downright ugly, I gave my Dada the doodle and sat hopeful, waiting for my sensational and life-changing Scooty to arrive.

The house would feel incomplete when he would go for work, it would be mute if it weren't for the chatter and chinwag of cricket commentators on his TV, discussing their views and unnecessary excitement of the match, and It would be an absolute scatter-house if his breezy presence were absent even for a second in the day. Forget one second, imagine living life knowing that one day, he went out for drinks and cards at his best friends house, and never came back home.

It was around 8:15 on a surprisingly quiet Saturday evening, I was seated with my sister and her best friend, cuddled and huddled up before the flat screen TV in my grandfather's room.

In that moment, we never would've guessed what was going to hit us.

Over the clamor and indistinct chatter on the TV, the phone began to ring uncontrollably and soon, the tranquility of that evening was overtaken by a monstrous wave of fear in its truest attire.

Fast forward to the next day, Next thing I know, I'm sitting in the waiting room of Max Hospital, with hopeful tears painted on my face along with a sheer glass layer of panic in my hazel eyes.

Ah, waiting rooms. It's absurd how one tiny room can have so many screaming souls, all tied together by the strength that each one contains. Waves of tumult and silent prayers enveloped the pale colored room. The smell of freshly brewed coffee mixed with the strong stench of hospital gloves, creates a balances in the room yet, a topsy-turvy ambience.

Hours we waited, hoping to get a call from the doctors of good news and discharge of my granddad, but……. nothing.

Morning turned to afternoon turned to evening tuned to late night and…..we were still there. The minutes that followed that long period of fearful waiting have slipped from my memory but all I can remember is seeing my otherwise strong and undefeatable father, sitting with his hands on his forehead, tears painting his face with awful colors of sad.

15th December 2013 a day that is engraved in my heart for all of eternity.

I remember vividly, there was chaos in the downstairs living room. There was indistinct chatter and rivers of tears.

From the railing of my staircase, I sat, clutching my knees, apprehensive to go downstairs and face the adversity that was being strung downstairs.

I sat and stared down and saw my grandfather, on a stretcher, being taken away from me, from this world.

At that moment, all I wanted to say was goodbye, but my throat was choked up as the tears were filling inside my body, waiting to explode. I couldn't hear my own voice, it was overshadowed by the rapid thumping of my screaming heart.

If I had that same opportunity, to draw a doodle or a statement of any one thing that I want most in this world, I would want to see my grandfather one last time to let him know that I will not let a scratch come to that Scooty I once wanted more than anything in the whole world because he gave me the whole world, not knowing that my whole world was him all along....

It's 3:18 AM, one again and here I sit in the very same gloominess of my very same room reminiscing.

You know how they say, "you want the things you can't have", I experience that every single day when I sit in disbelief in the darkness wondering, "where did all my angles disappear?"

The year 2012 was one that marked the beginning of an end in my family.

Around that year, my maternal grandmother the most intelligent and graceful woman I knew, whom we all lovingly called "nani", was undergoing a handful of medical negatives in her body.

She had had her pacemaker installed in the year 2008, long before anyone thought that that could have caused her a life of suffering and lifelong discomfort. Gradually, her pacemaker began to infect her heart immensely and soon had to be removed. The process of removal and reinstallation of the pacemaker caused further infections in her body.

When 2012 rolled around, Nani's condition had gotten worse and worse causing, my mother to spend most of my childhood, pre-double-digit years back in her hometown, Baroda, a humble city of intimate vibes and hospitable neighbors. Ah! "The great Baroda house!" The fond memories and uncontrollable laughs built the house that it is, till this very day. The characteristic smell of rustic softwood tables and freshly baked nutmeg cakes would fill the house, from all four corners. The sight of the Mughal stools by the window pane and the spirit of euphoria in the air made that humble Baroda house the best vacation home for my family and I. the pictures on the wall, I still have them in the back of my mind, engraved in my memory. The smiles that would greet us at the Sardar Vallabhbhai Patel International Airport could make me glassy eyed in the happiest possible way even today. That was the house that built me. I learnt the greatest lessons of life within the four walls of that tight and cozy abode, I spent majority of my childhood there wondering what it would be like to lose this house, hoping and crossing my fingers that I'd never have to face that and learning that those memories are what will get me through the toughest of days.

My mother would go back and forth to Baroda, staying there for months at a stretch. Hospital visits and crying at airports made my mother a stronger fighter than she already was. The tears she'd cried molded her into a fiercer fighter than she thought she could be. The battles she fought while she was away, only made her ready for the wars she was going to have to face, as were all of us.

Every flight she took, we would await her arrival with fingers crossed that the news she had brought would be nothing less than positive.

Gradually, her days in Baroda increased and the ailing state of my Nani soon became normalized as we had accepted the fact that she was very, very debilitated.

Bed-ridden and devoid of any forms of physical activity, she would lie still as can be, waiting for our worst nightmare to fish her away from us. She had come inches away from her deathbed a handful of times in the two years that she was at her weakest. Heavy medications and wired up to all sorts and types of life support, she lay there as still as can be. That was not the woman I knew. My Nani was a graduate from Stanford University who studied medicine and was one of the greatest in her field. She was always up for a challenge, however at that moment, she was fighting the most crucial challenge yet against herself, her toughest opponent.

The heavy medication which she was prescribed was of such mountainous volumes, that they were negatively influencing her mental state. "She seemed disoriented" I overheard my mother say, dispiritedly on the phone to a friend. As a young and dumb 10-year old, being disoriented meant nothing to be, merely because I was not aware of what it was and what it could do to the ones around a patient who had become disoriented. However, looking back at those times now, I wish I could have been more of a shoulder to cry on for my family.

As a child I remember vividly, my family was against the idea of me seeing my grandmother in the debilitated and frail state she was in, solely because they were afraid of how this would negatively impact me in those crucial stages when my growing up years were most impressionable.

It was around 2013 when my selfishness came into play. I did not realize this until a few weeks ago, when I was seated in my room and came across a bundle of notes, chits and letters my mother had written to me while she was away. The central idea of each note was to apologize to me for "not being around as much", for "not being present in my life when she should be", for missing PTM's, dance performances, birthdays, recitals and more insignificant events such as those that as of today, don't really mean much to me anymore. I was selfish when I believed that it was her obligation to be a part of those events, merely as an audience member rather than thinking of the dejected mental state she would've been in.

Yes, I was young but I was not heartless. Oh! How I wish I could turn back the clock and support my family more, be the light my family needed in all the darkness that clouded the usual brightness of their lives and to be the sympathetic daughter I could've been.

Now that she's gone I cherish every time I saw her smile, her frown, her childish tantrums or her hysterical laughs.

It's 3:18 AM, and though that might seem like an absurdly late time for most of us, the sane ones of course, for me it was only after the clock strikes three, I can be in the purest contact with my emotions. It is the hour where the heaven side of me identifies away from the hellish side of me, of my feelings.

It's 3:18 AM, and I lie here in the gloominess of my cold room, shocked and in disbelief, thinking- "damn I miss Nani."

You know how they say some things are just too hard to let go? (No one actually says that, it was just for dramatic emphasis) but, whoever thought of that was most definitely correct.

There are brief moments during the course of my day where I feel a sense of emptiness by just merely thinking of all the loss that my family and I have incurred.

Even though it has been six years since these events had occurred, there are instances where I deprive myself of sleep, and my thoughts of the memories of them keep me up for hours.

When going through a loss, you forget to breathe. You replace sleep with deep thinking and replace eat with over-thinking. Seconds feel like hours and days go by like years. Your sleep schedule intertwines with the hours you must remain awake and everything in your head is a topsy-turvy land.

During the moments where I'm not feeling that way, I feel a sense of pride for having known the three greatest people and a sense of relief that they aren't here to witness all the wrong doings and unlawful acts that are gradually deteriorating our country. There are days where I stop doing what I'm doing and just start to remember the days where I was entirely happy.

Although the concept of death is one that is truly hard to grasp and no matter how hard one tries, the feeling of absence of a loved one in your life never truly sinks in.

I was relatively young when I was faced with this adversity, and almost all my life I have been a rather secretive person who rarely feels comfortable with expressing her emotions. Throughout the phase of losing my loved ones, I never truly expressed my most authentic feelings about the loss I incurred. Not one tear I cried in the presence of my family, not one speck of weakness I ever showed, simply because confronting my own emotions terrifies me immensely. My family on the other hand was very candid and unvarnished about their emotions. Each of them would break into a series of fervent tears at some point or another. Me, being the only one who would hide their emotions, would feel responsible to be strong for the rest of my family and to pull them up when they slip backwards.

Fear-II

"The worst part of grief is that you can't control it. The best we can do is try to let ourselves feel it when it comes, and let it go when we can."

– Merideth Grey

*"I don't know what I want.
So don't ask me, cause I'm still
Trying to figure it out,
Don't know what's down this road
I'm just walking
Trying to see through the rain coming down
Even though I'm not the only one who feels
The way I do
I'm alone
On my own.*

– Taylor Swift
A Place In This World.

Welcome to the teenage glimpses! This chapter focuses on various glimpses of all the aspects of a teenager's life, from the perspective of a teenager herself. We don't know who we are, what we want to be or even who we want to be and representing the teenage clan, I can proudly say WERE ALL IN THE VERY SAME BOAT. So, step on this rollercoaster with me while I take you for a tour inside my brain.

Whether it's the pre-programmed desire to rebel against our parent's curfew laws or the passionate need to be locked in our rooms in isolation from the rest of the world. Being a teenager is hard. At the mere age of just fifteen, I don't know much, most things really, but I feel the need to just pretend like I do and put up a face of a wise demeanor making it seem like I have everything together (hint: yeah I'm pretty much dying inside).

Having been brought up with two elder sisters of inestimable talent of good character I learned a great amount from them, more than I ever would from any formal education environment.

They taught me right from wrong and never put me down. I was given high fives for the times I did them proud and something right, but for the times I caused a blemish to my character, I was sat down and explained to the reason for my stupidity.

All in all, I learned a great deal from both my sisters.

"I'm literally NEVER gonna be the way you two are, what the he-"

"Shut up Sam you're so dramatic" Trisha replied to me, staring me down from above the big cardboard box of cornflakes that divided her from me, that quite subtly deciphered that we really are quite not the same.

I sighed a sharp and high pitched sigh and reiterated my former statement; "but I wish I was kidding. I mean you and Aanch really have your shit together, I'm just kinda um, kinda like dying-ish??"

She chuckled and rolled her eyes at me and continued to chug down her glass of abnormally strong cold coffee, amused by my stupidity.

Oh I wish I was being 'dramatic'.

I wish I didn't mean the words I said but the sad truth of the matter is that I really don't know.

Being the youngest, I constantly admired and looked up to both my sisters in awe of one day reaching the level of the caliber at which they both were. Everything that they experienced, I experienced with them, from a distance, of course, merely as an audience.

From heartbreaks and heartaches, friendships and relationships, highs of the highs and the lows of the low, I was always watching from the sidelines, picking up all the little things from every situation that they were in, keeping in mind that one day I myself will have to undergo these exact encounters.

One thing about my sisters I truly admire is that, when they find something that they truly are passionate about, they will make a silent oath to themselves to put in all the hours of perseverance and fervency in order to excel in whichever field they wish to pursue. Their hearts would swarm with exuberance and zeal, to make their dreams a reality. They would work tirelessly for hours at a stretch, solely to outrank the person that they were the previous day.

Having a national level swimmer who excels in her studies along with a national level golfer who's everyone's favorite, it's nerve-wracking to even try and imagine a future like that for me.

Though it may seem like an undemanding and fairly simple task to find something you're passionate about, that is completely and utterly false.

Growing up, I wanted to conquer the world. I wanted to be every single possible thing. From a doctor (with the passion to recite stories at the near-by library) to a teacher (with a strong desire to take care of animals at the animal day-care across her apartment) to a ballerina (and a journalist on the side who bakes the world's best caramel cookie dough) to a musician (with a secret passion to design automobiles), there wasn't a single profession I didn't want, and trust me when I say this' I really gave that a lot of thought. However, over time, the fervency and enthusiasm that once lived in me heart and would light up my hopeful eyes, has now faded into thin air.

I no longer know what I want, who I want to be, where I want to go and simply what the need to know all these things even is. As a teenager, I have experienced a great deal of nothing in comparison to the older and (definitely) wiser generation. However, I do believe that the things I have learned and seen really have impacted my mental health greatly, leading me to have one motive in my life and one motive only- to be happy. Wherever I am, wherever I go and whoever I am, I have taken an oath to myself to let go of the toxicity that once contaminated my euphoric spirit and to find the good in all situations, whatsoever.

This section is a glimpse into what a typical teenager looks like while juggling the never-ending academic pressure while simultaneously balancing their own emotions and personal problems.

(We teenagers belong in a circus act.)

The bags under my eyes look painted on as if I switched lives with the clown who scares most of the little kids away.

The usual radiant smile on my face has molded into a look of perturbed horror, kind of exactly like that clown who changes his expressions to fool an audience.

My otherwise jubilant and bubbly persona has now sculpted into a blank canvas that refuses to get painted on, similar to that stubborn lion at the circus who REFUSES to take orders from it's persevering ring master.

I had changed.

As teenagers, we each go through a handful of changes in our bodies be it physical, hormonal, mental or emotional, a teenager is a true work in progress.

On a daily basis, we each encounter various different situations where our emotional, physical and mental capacity is tested simultaneously thereby draining us of all our energy.

Now, when the exam season approaches and makes its presence known, our ability to balance our mental, emotional and physical strength is truly put to the test, while we now have to juggle our academics along with that. Exams are and have been a part of almost everyone's lives, therefore the crazy stress of it all I haven't quite comprehended.

However, when I sit here typing this chapter really thinking and analyzing what really causes this crazy stress about exams, I begin to mentally list down all the external factors that make the exam season numerous times worse.

We as the youth of today's rapidly evolving modern generation are given various tags and labels that assume we are something that we are not, someone we are not. With the changing times, the ones around us, with the generational difference, expect us to excel in every field we try our hand in.

However, what they don't quite understand is that with the negative influence of media and the judgmental generation that we ourselves are, it is exceptionally difficult to even feel accomplished anymore.

With unmerciful comparisons and abnormally high expectations, it is hard even to merely breathe in the toxic environment that surrounds us.

When it comes to exams, the pressure is intense. I can almost say it's a teenage war zone. It is especially nerve-wracking because, we don't only want to get a reasonably good final grade, we want to beat our own ego that falsely tells us how much superior we are to everyone else.

Exam season takes a toll on our mental and our physical health.

We work tirelessly throughout the day and late into the darkness of the night, causing sleepless nights that result in extra groggy mornings and a feeling of immense fatigue and exhaustion from the previous night coupled with the mindset that we have to do the exact same thing once again and then again and again and again and again and again and again and again and again and aga-

Though the disadvantages of exam season outweigh its advantages by a GREAT margin, the one thing I find particularly interesting is, just how much we discover about our own selves in the process.

Exams bring out the very best in us. We are able to recognize our inner strength and will power to do something with so much precision. We are able to recognize our ability to tackle our raging emotions and personal problems with the academic stress and strike a balance between the two, with so much ease.

Things That Helped Me Tackle Exams

1. Realizing that studying with your phone around is not really studying…..
2. Replacing unhealthy snacks with healthier alternatives

3. Taking frequent breaks
4. Studying in chunks
5. Dividing your workload among 3-4 days so as to decrease the exhaustion you feel if you study all the material in one single day.
6. Watching helpful Youtube videos related to the topics I'm studying.

To sum this section up, remember that getting a good grade or a bad one does not define who you truly are (I'm sorry I know this is probably the most clichéd thing I can say but eh). When giving exams, we need to keep in mind that making unhealthy and unnecessary comparisons with your peers/siblings will not benefit you in any way whatsoever. If you're anything like me, it will demotivate you and push you to a point of not working hard enough because you believe that you won't reach the level that those peers/siblings are at.

I don't like sounding like a preacher but it is so immensely important to not get swayed and buried under the literal burden of your books and be dug so deep into a hole of stress, worry, doubt and high expectations that you are not able to climb out of it, even after exam season has ceased.

Stay calm and breathe because we all make it out alive at the end of the day…..

I won't label this section anything. Simply because this topic has so many labels attached to it as it is, I might as well leave it blank.

> *"People on the airwaves always preach who I should be*
> *How to dress*
> *How to talk baby how to breathe*
> *Wicked artists use their limber fingers*
> *To paint pictures of what young girls*
> *are to aspire to be"*
>
> – Catie Turner
> 21st Century Machine.

She stares at herself in the mirror analyzing every blemish on her body. She adjusts the shoulders of her ill-fitted V-neck T-shirt and pulls up her black jeggings, in an attempt to look presentable for society as she gets ready to step into the teenage battlefield. She pulls her hair back into a roughed out high ponytail and ties a tight knot with the elastic around her wrist. Shortly after that, she rips off the said elastic and begins to comb her hair all.over.again.

Her face: painted with not only Maybelline's new and improved shade 345 matte finish foundation but with a subtly evident tinge of discontent, splashed all over her face.

Her eyes: Glassy and puffed up from the tears she had just cried over her dissatisfaction of her own appearance.

Her arms: "fat"

Her legs: "ugly"

Her smile: missing.

I forced a smile and painted my lips with the desired colors of acceptance and made my way to the teenage war zone…I feel the need to rush to the washroom immediately and alter my appearance…like NOW.

I skipped a hundred hellos on my way, as I panickily made my way to the NEAREST washroom.

I maniacally wash my hands, believing that that would solve my problems.

I stare into the mirror as I hurriedly search for a tissue to wipe off my insecurities.

I stare deeper into the very same mirror, with a bit of a bewildered horror and rip out my perfectly tied ponytail, once freaking again.

"You look disgusting, did you really think your stupid ponytail looked good???"

"Ha delusional idiot"

"She really thinks she looks good, should we break it to her??"

"Those are so not the right sized jeans she looks so fat HA"…said my inner voice.

It is not an external factor that causes me to feel this way.

It is merely my bitchy, insecure and probably pretty honest inner voice pushing me to believe that I am not happy about the way I look….

"Hey Sam!!! DUDE you NEED to try out for this part", I hear someone shriek as I pass by the beloved lunch hall as I make my way to it hurriedly.

I paste a fake smile on my face as I greet this random stranger who seems to be on a nickname-calling basis with me and weirdly knows that I am or have been pretty in love with theatre and knows that yes indeed I will be great for that part.

She waves a flyer in my face and insists on giving it to me.

She walks away briskly as I stand there admiring the big font that reads "TRY OUT NOW THEATRE ENTHUSIASTS, DELHI'S BIGGEST OPPURTUNITY!!"

If you had given this exact flyer to me about a year ago, I'd already be there, standing in line waiting for my name-call.

I love center-stage. I love the sound effects in the background, the cheering audience, the uncomfortable costumes, the over the top makeup and above all, the feeling of conquering the world, at center-stage. However, growing up and reaching my teen years, that sure did change quite a bit.

I started holding myself back when I would hear about opportunities to act on stage. I would get insecure and self-conscious at a simple task like reading an excerpt from a lesson during an English class. It reached a point where I would literally skip a class, just to avoid answering a question or read an excerpt out loud.

My self-consciousness had reached such a great height that it had shifted me to a state of shock and disbelief. It became so drastic to a point that I stopped going out to social gatherings only because I hated my own appearance. I hated putting on a pair of ill-fitted skinny jeans and interacting with people with whom I would feel especially self-conscious around.

I had identified this as an inherent issue I had developed and was afraid of who I was becoming and that I would not be able to bring my old self back.

My friend called me one night, at an oddly late hour and sounded exceptionally panicked on the phone.

When a friend calls you in the wee hours of the night, you tend to pick-up, thinking woah this is an emergency for sure.

I picked-up in a literal second and skipped the formality of "hello" and "what's up?" and straight up askd her what the HELL happened???

She sounded unnecessarily panicked as she spoke and told me that "her birthday dress isn't going to deliver in time for her party" and then she cursed Forever21 for a few (long) seconds to blow off some steam and then finally SHUT UP.

I rolled my eyes (although she couldn't see that) and said in an exasperated and obviously sleepy tone "bro it doesn't matter go to sleep its three in the freaking morning" and briskly hung up the phone, slamming it on my bed side table, knocking over a thing or two.

She was obviously taken aback by my response and so, decided to drop in a couple concerned texts that I saw only when I woke up.

They read:

"Uh"

"That was pretty odd"

"When did u become so serious??"

"lol this is so not u"

"you're not the same anymore bro"

"I'm not overreacting I swear I've been meaning to tell u this"

"anyway I'm gonna crash now."

That's the one that pushed me to re-think the decisions I have been making.

It's not about that 37 seconds long phone call I had at three in the morning or the multiple texts that blew up my phone shortly after that concerned me; it's the fact that I have known subconsciously for a few months now that I have changed a hell of a lot and the fact that it has been so difficult for me to transition back into being the girl I was before a wave of self-doubt drowned me into an ocean of insecurities.

Coming from someone who is insecure and self-conscious (to a certain extent), there is so much rage and anger that fills your lungs, making it hard to breathe that you tend to exhale that fury in the form of ugly words that we don't mean that we regret later as we lie awake at odd hours, regretful and embarrassed.

This section was the hardest one for me to write solely because it is the one section that deserves to be the best one as it revolves around the two people in my life who I hold most important.

> *"Caterpillar in the tree*
> *How you wonder who you'll be*
> *Can't go far, but you can always dream*
> *Wish you may, and wish you might*
> *Don't you worry, hold on tight*
> *I promise you there will come a day*
> *Butterfly fly away"*
>
> – Billy Ray Cyrus
> Butterfly Fly Away.

7 years old, everything was right. Things were easy and life was uncomplicated. The relationship I had with my parents was at an all-time high. When you're young, most of the people around you seem invisible. But then, there are a few (maybe 3-4) people who seem to stand out. For me, the two people happened to by my mom and my dad(and of course, my two sisters).

Growing up, I had a fairly solid relationship with my parents. At eight years old, there wasn't much to discuss anyway. They knew about all my friends and the fights we had had, the memories we shared, the classes we had, the food we ate, and all the other insignificant minuscule details of my days.

Having two older sisters who were rather frank about their feelings and affirmations, I was taught early on to communicate your emotions and to never let things bottle up. However, as I reached my teenage years it consistently became a more difficult challenge for me to approach.

My argument, however, was always that I know what I'm doing and know which path I am on. Tie me up to a lie detector machine and hear it make the loudest beep. That is utter rubbish. I never know if the decisions I am making for myself are the most beneficial ones. What if my friends aren't as genuine as I thought.

What if this boy I like is not who I think he is. As a teenager, none of us know what we are doing yet we ALWAYS ALWAYS ALWAYS put up a face of wise disposition and merely pretend as we do.

There are days where I gather up the courage to attempt to talk to my parents about anything that's bothering me but, I always find a reason to scare myself away from that possibility.

No one really realizes the amount that I want to communicate my feelings but it is just so completely difficult. It is so alien to me!!!!

By age fourteen-fifteen, I had so much on my mind that I just needed to get it off my chest. And so, I found comfort in my mom and decided one random Thursday night to tell her over a take-out Chinese dinner, all about my "secretive" teenage life I've apparently been living.

It was surprisingly hard but it had to happen and I'm glad that it did.

My dad, on the other hand, is a tough nut to crack. I remember being nine-ten years old and spending almost all my time with him. We had a fixed weekly routine. Every Saturday we would drive down to his site, he would show me around and would explain what was going on. Shortly after that, we would go to the mall right behind it and go to one specific store. He would always pick out something for me. To end the day, we would walk and browse around and finally get breakfast. That is how a Saturday morning would look for me, every single week.

But now when I look back and reminisce, it hurts me to admit just how much things have changed as I've grown up.

As teenagers grow up, we often tend to unintentionally detach ourselves from our parents and tend to push them away. Even though it is purely involuntary, it happens it really does. It is one of the hardest things to go through not just for us but equally for the ones we tend to push away.

We feel like the world is against us and no one can hear our screaming voices. It is a frustrating feeling. You feel trapped somewhere you don't belong and can't get out.

With time, however, this feeling evaporates into thin air once we come to terms with the fact that the world does not revolve around us and the people around us whom we push away are indeed not trying to hurt us.

As a fellow teenager, I feel authorized to make this PSA:

DO NOT push your parents away from you. The limited time you have left with them, please do cherish every second of it. They are not trying to hurt you in any way. They mean the best of the best for you. If you've come this far in this section, use this is a sign from God to send a simple text to your parents, just asking them how they are, I know I felt good when I did:)

This section is an ode to someone who left an imprint in my heart, in my life. You were there once and in some secret and subtle way, you'll always be there.

> *"Who are you?*
> *You are not the girl I used to know,*
> *You had my back when I would go,*
> *Be there through high and low*
> *Who are you?"*

> -Tate McRae
> Hard To Find.

In life, you meet all sorts of people.

Some that stay, some that leave, some that frequently leave but come back eventually. The one thing however that each of these people holds common is that each of them leaves an impact. The impact may be negative or positive but there is always something.

Losing this one specific person took a toll on my mental, physical and emotional well-being. I remember getting zero hours of sleep and tumbling into school looking like the walking dead with dark circles that looked painted on.

She was the only person who knew me better than I knew myself.

She knew the stories before I could even begin telling them, she identified my lies before I even looked guilty, she had my back when I wasn't myself and she was always there to put a smile on my face. I knew her inside out. I knew her strengths that made her who she was and all her weaknesses that made her flawed in the most beautiful way.

She had become such an important part of my everyday routine that not a day would pass without hearing her voice or hearing about all the small insignificant details of her day. She knew all my secrets and swore to keep them locked from the rest of the world, forever and I believed it when she said I could trust her.

However, all good things must come to an end and people can't stay forever. I learned that the hardest way.

It was as though I woke up one morning suffocated under the weight of a thousand rocks, suffocated under the bitterness of the truth.

I woke up knowing that she was no longer a part of my life anymore and was forced to shove this sour truth down my throat.

That was literally rock bottom.

There was no place for my tears anymore.

She was the hand that wiped the tears off my face, but in this topsy-turvy reality that was my life, I cried those tears for her.

Seeing her every day and knowing that she is now just a stranger with all my secrets was the toughest pill to swallow.

I still reminisce the days where we climbed every mountain, where we rode every storm, where we tackled every fighter, where we danced to every song, sang every song and did all of this together.

I still sit in disbelief, glassy-eyed, trying to wake myself up from this nightmare.

Oftentimes I still try pushing the rocks off my chest but no matter how much strength I put into doing that, it is never enough. It will never be enough.

It shocks me to see that a part of me is still missing. Avoid unfilled.

A heart still left broken.

It shocks me to see how far we've come, how much things have changed.

It truly blows my mind……

An ode to the ones who cry themselves to sleep every night, worried and afraid about who they are, who they want to be and what they're doing. This one's an ode to the ones who feel lost, stranded on an island of tumult with no way out. This is for the ones who have felt like the whole world is against them. This one is for people like me, who are still trying to find their place in this world. And you know what the best part is?? That is perfectly okay.

> *"I am trying to learn and I'm dying to know*
> *When to move on and when to let it go*
> *A curious feeling no one can explain*
> *And I just don't know if I'll risk it again*
> *When the future's so unsure*
> *When the future's so unclear"*

– Kodaline
Unclear.

Sometimes, the words don't slip past your tongue, even when they're aching to fight their way past your lips, where they're placed. Sometimes you don't have the words to express how you feel. Sometimes a song can do that for you. Songs are my voice when the words don't come out of my mouth. I have mastered the art of not knowing what I am, what I want to do or even who I want to be. But that's okay because just like everyone else, I'm just a girl who's trying to find a place in this world.

Fear-III

Acceptance

"I don't know what I want, so don't ask me
'Cause I'm still trying to figure it out
Don't know what's down this road
I'm just walking
Trying to see through the rain coming down
Even though I'm not the only one, who feels
The way I do
I'm alone, on my own
And that's all I know,
I'll be strong, I'll be wrong
Oh, but life goes on
Oh, I'm just a girl
Trying to find a place in this world
Got the radio on, my old blue jeans
And I'm wearing my heart on my sleeve
Feeling lucky today, got the sunshine
Could you tell me what more do I need?

And tomorrow's just a mystery, oh yeah
But that's okay
I'm alone, on my own
And that's all I know,
I'll be strong, I'll be wrong
Oh, but life goes on
Oh, I'm just a girl
Trying to find a place in this world
Maybe I'm just a girl on a mission
But I'm ready to fly
I'm alone, on my own
And that's all I know
Oh, I'll be strong, I'll be wrong
Oh, but life goes on
Oh, I'm alone, on my own
And that's all I know
Oh, I'm just a girl
Trying to find a place in this world…

– Taylor Swift
A Place In This World
The Beginning Of The End-
'Dancing In The Rain'

"There's an end to every storm. Once all the trees have been uprooted. Once all the houses have been ripped apart. The wind will hush, the clouds will part, the rain will stop, the sky will clear in an instant. But

only then, in those quiet moments after the storm, do we learn who was strong enough to survive it."

– Meredith Grey

Acceptance is a flexible word; it means a million different things to a million different people. To some it may mean comfort, to others it may mean a way of life. However, to me acceptance means to conclude with greater wisdom.

To accept what has been done, to let go of the toxicity that surrounded a certain situation and to move on from the pain that was latched on to you like a growing flower, glued to the dirt in the ground, refusing to move an inch, acceptance has been tough (to say the least).

I remember when I first started writing the first chapter of this book, I was SO immensely in touch with my emotions that I would cry at every draft I would alter even the slightest bit because I was constantly reminded of all the pain I had been through to get myself to write that chapter. I remember sitting in my room with teardrops flooding my cheeks, puffing my eyes all painted on the space bar of my laptop, spreading across the bottom half of it. I remember every word of every sentence of every situation that brought me to write my fraction of this book, because the story I told was the one that helped me conclude with "greater wisdom". However, one thing I never expected from myself was to finally accept that all of those instances brought me even closer to reaching the final stage of grief, acceptance; the conclusion to my story.

Acceptance does not mean "to forget" or to "erase from memory" as some great act of strength or bravery. Acceptance means to comprehend and move on from the grief that once surrounded your aura and to realize that it cannot be your whole world, and can only be just a fraction of it.

The first battle I fought when I was inching towards acceptance was to eliminate the fakery from my life; removing the "masks" that so skillfully hid the truth behind the fake smile and the bright eyes.

The battle was to accept the frown and have it exposed to society, unfazed.

When the truth is out there, raw in front of the whole world to see, it becomes easier to adapt to the fact that fake happiness is a mirage that hides the truth, and more often than not causes an even greater deal of pain.

The question that lingers at the back of my head, poking out in the wee hours of the night and that keeps me wide-awake is, how do I remove the mask? The mask that is my superhero cape, the mask that protects me from being too vulnerable, the mask that gives me the strength to fight, the mask that has become a part of me, the mask that is the person who I want to be. How do I remove that mask?

The answer lies within us. It lies in the blood, bones and skin that make our bodies, functional bodies.

Majority of acceptance is accepting the person that you are and not who you want to be simply because, the person we want to be is hidden inside the person that we are and until we accept our own self, that person will continue to hide, making unfair comparisons, putting you through hell.

Removing the mask means accepting the fact that it must rain and pour and thunder until we have a peek at the rainbow on the flip side.

We remove the mask by understanding the greatness that suffers silently underneath it and learning the importance of exposing that greatness to the world and to ourselves.

The mask is a shield. The mask is protection. The mask is an umbrella. An umbrella? Have you never danced in the rain?

After standing under the pouring rain, breathing in the tranquility of the grayness of the clouds, after inhaling the sweet scented monsoon air and filling my lungs with that greatness, I never carried an umbrella when it rained; not even once.

I danced. I danced to the sound of the drumming of the raindrops against my window-pane and sang the song of the thundering skies.

I danced in the rain and my mask gradually soaked up the rainwater and slipped right off.

That's how I removed my mask.

Reconstruction

– Nandini Shukla

Reconstruction-I

"We live in an era of do or die. We have constant "hows", "whats" and "whys". Humans are imperfect, that's quite evident, however, nobody can limit us from being near-perfect." I understood at a very young age that neither life is fair nor one always get what one wants. Frankly considering the generalized ideology, everyone has to go through their nebulous turns on the road of their lives, while some may choose the road least traveled and some the most walked on. Everybody gets their share of happiness and regret, which we can never neglect.

The simplicity these thoughts above possess is something to be embraced but the depth from where they arise needs to be highlighted. We all have an individuality and I do like the idiosyncrasy I hold outwardly and within. Somehow there still lies secrets hidden inside, hurt is still faced presently by the past sins. Success is achieved in intervals after the failure that spins. I keep asking, till when do I have to cover up my mistakes and end the day with a win. I am Kiara, 17 years old creature and pretty positively messed up yet an ardent sophomore who has a fear of not living the life she dreams of. An achiever in studies, fairly popular in school, co-curricular at an A-grade level with sorted relations with people (ignoring the haters). I find myself sexy and sometimes I call myself pretty when I feel it, haha. Moreover the comfort I have in my skin and the way I love myself profusely has not come very easily. It took me a considerable time to adore the way I am without a barrier for further enhancement in my learning. It sounds philosophical but I am not quoting anything because maturity doesn't come by age, you just develop that particular attitude by experience. The generation we live in has boosted the standards of living so much that most, not all of the adolescents gain the knowledge of pretty much so-called 'life' quite early, which is both, bloom and doom. Likewise, I am different. The life I started with was not very conventional. Born to unsettled parents with one against the very idea of having the 'baby' was one drawback which made me have a kickstart. Despite being the only child, I wasn't pampered at all. A small town never gave me the joy of growing up with nature and playing around near the lakes or gardens in the afternoon. The imagination of having wooden toys or the toys associated with girls to play with, while munching to favorite snacks made by grandmother, having evening walks till the ice cream vendor with grandpa, swings being the deadliest yet the most exciting places, was only a fantasy which was never a reality for me as a child. I grew up learning the basics of human's everyday activities of walking, talking, eating, etc., on my own with the help of my five senses. Yes, I feel blessed to have all my senses working thoroughly. The day the elders in my family saw me being a little independent, the little me who just started practicing to be on the toes without tripping often was shown the gates of a cheap boarding school because parents chose career and grandparents couldn't take the burden anymore. It wasn't that they couldn't afford me, they just felt that raising the child was a tough responsibility. They wanted to skip 'childhood' and then grab hold of me so that the time they would have otherwise spent on cleaning my stuff, teaching me common things and taking care of the tiny creature's further expenses could be eliminated from their lives. I wouldn't blame my parents or my family for not giving me a single childhood memory expect how the absence of their love, their affection, warmth and their very presence from my life made me turn out to be a not very pleasant child. The boarding school I was in did deserve the title of being 'cheap'. With not a lot of staff to look at children and lack of resources to keep up with their needs, knowing that my parents were alive I still had a taste of how an orphan feels. I got spoilt. The very memory of leaving the premises usually at early night just to roam on the roads and rob the fruit vendors to eat the juicy fruits we could dream in the day and learning to sharpen various tools like broken knives and blades etc (what we considered amusing to play with) with other children who knew much more things than me and having fights which ended up me being covered with bruises is a memory that I can never burry. After five years of my initial stage of life, the boarding school taught me enough things to hold on to. I wanted to grow to be a strong and tough, docile and well-mannered child but I had other attributes which were complimentary in me. As the little Kiara, I could be defined as rough, arrogant, selfish maybe, abusive, stubborn, a person who used to not interact much with people because of her tendency to have hate at first sight. A fake smile was always there on her face, structured lies she could tell to run away from home to be with her unfaithful friends. Forget watching movies and playing with barbies, I was more interested in knowing how to unlock a car door and cut pockets neatly, because I had the most 'talented' company in the weirdest area I would say, where my parents bought a small house with two chairs and a handful of utensils and a sofa-cum bed. I joined a private school by a decent age of 7, just because they wanted me to be an asset to them not a liability. I can thank them for something at least. However, the personality I had turned myself into as a child gave me a terrible time in school. Least friends I had and that too because of the fake persona I could portray. Little I could do to understand what I was learning and teachers seemed stupid to me. The school wasn't fun because I missed my dirt. Going back home was okay till the evening used to come as my mother and father had contrasting personalities and both of them wanted something else from their lives, which made them fight like cats and dogs and me being disregarded was just a consequence of their mistake. But things started reshaping, because of school I could pay little attention to my so-called 'friends' in the perilous colony. Dad had to shift to Africa to get ahead with his career and money because he had a soft corner for me, and somehow wanted me somewhere to be. Mother was that parent, who was against me, she could never 'love' me because of all the things my birth made her compromise. Sometimes we often get our enemies disguised as people in our family and our family is found in people, we could least expect. That's just one of the ways life makes a human strong enough. I was never a victim of bullying in school, but I was well experienced with home abuse and victim of my mother's terrible intolerance towards her anger. I did leave fighting but fighting didn't leave me. My little mistakes used to provoke the lady in my house to hit the child as if I was doing a crime. She convinced me well that she never wanted a baby. Her heart didn't and can never lie within me. I developed the habit of reading and as much I exposed myself to the world that laid outside the four walls of the room and my school, there weren't restrictions on what as a child I was supposed to read and therefore I came a long way in enhancing my knowledge about the things I experienced and I wanted to be exposed to. The journey wasn't all that shady, I just assume it to be an adventurous past story, not full of glory yet an interesting misery which can be perceived as a brave history. Haha. Well, I do embrace the fact that from a young age I knew how to cook, ran my errands and do most of the things on my own. Little did I needed my parents to look after me. Gradually scoring good marks in school, started becoming one of my priorities. Being into sports wasn't an onerous thing for me, I had a tough body as a child and was very athletic than most of everybody. I started developing an attitude of empathy and kindness towards people because I started realizing how it felt when somebody for no reason replied disrespectfully. I became a deep person because of the hurt I encountered and my past taught me many things beforehand. I can feel others when they get hurt and upset. I understand people in a way they want to be understood because I know how to practically deal with emotions. People confide in me for a reason. Home issues are my reality from which I can't turn my way around, but stubbornness never died. I got firm on the thought that I wanted to excel and wanted to be more than just a child who knows how to study the school-published books and do some extra-curricular. I wanted to redefine the 'different' in the different life I was living. It's very likely to see people become those who they were taught to be, but very unlikely when someone turns out to be something which he/she could never predict to be. Life has a peculiar way of teaching people to find their purpose and justify it. We are not meant to walk in the biblical 'Garden of Eden'. We walk through our rough paths. If not then, then now; if not now, then later. I can reach beyond my imaginary lines of limitations similarly I am subjected to learn way more than anyone. Growing up, I did have relations with people other than my parents. Some souls left, some stayed and there are souls I am eagerly waiting to meet. Those who left taught me lessons, but gave me internal scars. It's not as easy as it seems when somebody says, 'move on, people come and go!' it's true, but the truth is sad. I still look back to the way I was treated negatively in a different way, with being used and neglected. I was shown one more side of the world. Those who are staying, have my love, care, attention, and admiration. They are considered a blessing. It's because of them I don't preach, 'the world is cruel', because of them I have hopes that positive emotions do exist. Those that I am eagerly waiting to meet, I don't expect anything rather I am glad that I will be more observant when it will come to including anybody in my circle. Today, I do get drowned in my flashbacks. I still am short of clarity about my reality. I have this urge to make my truth and not accept what is true. That feeling of wanting more than what I do is so prevalent yet not relevant. The kind of disorder I today go through is not something which can be cured with some prescribed medicines. Those inevitable emotions that I feel suddenly out of blue, about which I have no clue. I want to expound my noble-order, however, I don't know where to create a border.

Sundown, yellow moon. I replay the past. I know every scene by heart. It all went by so fast. I live on the mercy of me. That urge to establish your dignity after humility.

Being misunderstood, judged and taken for granted, I think each one us has somehow in some way experienced it. The 'want' to change the fake smiles to sincere smiles. To finish the crass conflicts with people I don't like. I don't want to be stuck. I genuinely don't. Helpless? I don't want this word to be in my dictionary either. I don't. I don't want to compromise on my differences and fake similarities to adjust in the crowd just because I don't want to be left out. I have trust in my idiosyncratic reality and I want to be my person and be loved for the way I will groom myself. But I don't only want to 'want' all of this. I know as I told before, one doesn't get always what one wants. I can't take that risk. I need this. It's my life after all. I crave the best. I want to achieve, I want to rise with shine. I want something out of this life of mine.

Reconstruction-II

"Kiaraaaaaa! Go wash your face! The chapter isn't even that boring to make you fall in such a deep sleep." said, actually shouted the social science teacher. Day 3, when I fell asleep in my classroom in the subject period which used to enthusiast me once. I keep wondering what makes me so tired for no reason, not that I get physically tired but the brain does get exhausted and demands rest at odd times. I sleep most of the time in school, genuinely argue with most of the teachers I interact and I have good friends to support me but there are a lot of other people, 'girls' specifically who dislike and would love to throw me off the building due to their insecure nature about their so-called 'guys' getting distracted from them over-me. Usually, when I enter my classroom, people just glare at me as if I am from a whole new world and have come to capture people in my charms! In general, when somebody(me) disagrees to open up about themselves to others and is extrovert yet keep their lives private, people love to cook up recipes about them without knowing how bitter they can taste in the end. I exhibit a candid hatred towards school, not that I have ever been suspended for no reason, I just don't like it. Its that kind of hatred, where one doesn't know the reason but still have no vibes to like it enough. You see, I kind of have rigid practically pessimistic thoughts about general things in my life. Not that I don't appreciate things, frankly speaking, I am an admirer but when I have setbacks- all the bright colors of life turn into shades of black, not even grey, which makes me just a dull person emotionally. Olivia, my one girlfriend who has observed me the most and with whom I have spent a considerable time,(its true that we have been called lesbians) she knows me more than my other close people even if we aren't together anymore as people change and I don't expect them to stick with me till the day I die. She tells me that I think very highly of me. I get arrogant with people when I come to make a point clear to them which they weren't initially agreeing on with me. She also tells that I make harsh judgments more on myself than others and my sadness and pessimistic emotions hurt me deeply but somehow people can't figure out what I might be going through some deep shit because my grades manage to be intact and otherwise general performance remains unaffected. Most of my friends marvel at me and I do get repeated questions, 'Kiara how do you manage it?'. It was easy for me to wake up and tell myself that I had to fake it till I make it, not realizing the fact that by holding up to my emotions and burying the distractive thoughts can someday reach up to that maximum level and can make me burst out and leave me shaken and broken.

'I have a normal life. I go to school. I have the ambition to become something meaningful and successful in life. I come home. I like dancing and reading. I have pursued hobbies. I keep exploring things. I do homework. I study sincerely. I listen to what my mother has to say. I make a point to talk to dad alternate days. I am there for my friends or at least I try to be there for most of the time when they need me. I interact with people. I love somebody and that somebody loves me back, It seems to be a happy relationship. I go out and have fun. I cook breakfast for myself and have me-time every day before I sleep. Wow! Trust me this is what I would call as 'my myth of normality!'. I feel like laughing so hard till tears come out of my eyes and I want to keep laughing till those tears transform my laugh into my terrible cries.' I was discussing this the other day with Stel, (he is one person I can trust blindly. I am not just saying it to tell how close relation I have with him, but when I say it 'trust blindly', I mean it.) It takes a considerable time for me to gather all the sad stuff I have been dealing with recently and then finally having words to define my feelings to somebody, that is exactly why I behave odd at unusual times and make people wonder what happened to me suddenly. Haha. Anyways, so I was saying that whatever diagnostics I did on me and the satisfactory explanation I could come up with to explain my downhearted state, I needed to spit it all out without being judged and questioned too much. The only possible person who could just listen and then just provide comfort was Stel. He understands that sometimes people don't want any kind of unprofessional therapy or brief advice or long lectures on 'how to cope up' with their miserable state. Sometimes all one needs is a person who can lend his/her ear and pretend to be an empathetic statue on whom you can take all the burdensome words out and be normal again. I believe that not all of us have those metal issues or blocks which needs to be examined by mental healthcare experts, I think that when we fall short of clarities in our life, like when I suddenly stop walking and get lost in some thought or when I feel those kinds of emotions which make me sit back and make me tired enough mentally, in order to stop me from doing work or when I go through days where I get stuck and I fail to grow through... I think these situations of ours needs to be understood by us in the end, as we are the only people who have the complete authority of ourselves and we have to realize that no matter whomsoever we seek out for solutions to our problems, until and unless we remove that over-dependency on others, we will fail to dominate our very own life and not give ourselves the chance to even self-analyze and learn through the hardships we are facing because I do not know much about others but when I see my people understanding my state, that's okay! I feel happy that I have people there for me and it's alright for me to listen to their perspectives on my issues but I get frustrated when I see that they don't feel what I am feeling at the moment. I can't even play the blame-game. All I can do is care for myself but then again, I feel and feel like I am caught in a cycle of unwanted mistakes that I keep committing which makes think of myself from inside that I'm a bad person. I have experienced no success when people tell me how I can make my condition better because until I give myself my therapy with my self-analyzed theories, I find it hard to move on. There are repeated questions I am asking today to myself- why I do I feel the way I felt?

Why is that I sometimes can't define my sad state? How will I change for good? Where am I exactly stuck at? Or who am I stuck with? …

Well, I have always that urge to define my emotions to remain as sorted as possible, but while knowing them I forget how each emotion feels actually. I do not know much about the world, but I when I get connected with someone, I overdo the feeling –part and when these connections of mine, become some prominent figures in my life, I fail to equally prioritize them, it's ironic to see that I was the one who brought them close and then due to my circumstances and some of my actions which I fail to control, becomes the reason for their disappearance from my life. So, as you can see, my frustration is how I let people go after I get tired of manipulating them for staying in my life, when in the first place, I do not even know why I want them to stay, I just feel that I love them and I feel that they should stay. My hurt is basically due to how I love myself and the company that I want with myself because I am obsessed with the growth of my persona, however, I let myself get completely away from the people that expect something of me, people that love me, and people that consider me something in their lives. People think I have a way of attracting anybody that I meet, It is true that when it comes to interaction with somebody I am meeting for the first time, I will always put the best version of me. Never have I ever messed up the first impressions and I have these instincts which do not let me go near the people I know with whom I won't have any bases for interactions and I would barely be able to carry on with conversations. When I look at someone, if I am able to get a positive vibe, I can make the person talk to me for hours. My happiness can reside in that person for as long as I want. The problem lies when I am not able to be consistent in giving the time they expect, the attention they want, and in demanding all of this, I believe they are not wrong, as it was me in the first place who told them, that "if they weren't special, I wouldn't have made efforts to try keeping them in my life at all". Joe, a person who is in my school and we started talking a lot in the years when are classes got the same. Initially, my taste didn't lie with him, because I was too busy trying to make a connection with somebody I always wanted to date since the time I first started dating. Haha. But gradually I figured out that joe, was a good person by heart and he came in those few people with genuine intentions, moreover his loyalty as a friend was something to be applauded, however, his anger was an issue which is okay, I don't really mind that because I have experienced worse issues with people. We got close and he became one of my really close friends. We used to hang out and his family was one of my favorites, I loved being these people and everything was amazing with him, but soon I got stuck in the cycle of 'things to do to make you ready for being an ivy league graduate' and 'things to do to make you a personal asset', that I just couldn't give five minutes of the day to any of the friends who were dear to me. It's not that I have a big circle. Within a small radius, I have 2-3 people who know my history, my ambitions about future, but what upsets me the most is that I give up in making them truly understand what actually I am doing in present and what really drives me crazy, (although, it shouldn't), is that I can't explain in details about what am I doing because, to be honest, I want people to know that I am good at it, not how I got good at it. I want people to see my final destination, not the path I walked to reach there. But when I constantly keep pushing plans with away, without justifying what's more important, it's obvious for them to think that I took them for granted. That's why I love just being with me, it's not that I have a fear of people, I can be with as many as I want, it's just that I don't want to deal with misunderstanding, that's when the point usually comes in to keep the ones who understand in few terms and still manage to love and say goodbyes to the ones who can't comprehend my state, no matter how close they were. I am not a heartless person, I do try to keep up with people I love but sometimes I just remember the good times I had with them and due to the universal law that states 'people change', I manage to be merry in the memories I had with them and letting them go doesn't seem a big deal to me, I just will miss them but soon the feeling will fade. The situations hurt me most, when I try to be as genuine as possible with people but somehow people fail to embrace imperfections and expect the others to love the idea of accepting everything as it is. I try hard to always be my best version with people I love. I don't know if that's bad or good, however, I know that's what I do. When I come to truly be me, often I disappoint others. These energies to making me live up to the expectations of my loved ones and to always be the reason for their smile, sometimes trigger the six stages of grief. Of course, we are imperfect and improbable. We love preaching, "love yourself the way you are", "embrace the emptiness that you possess", and all other self profiling prophecies, but when it comes to appreciating the flaws, we fail. We fail. I fail. It can't be neglected, that we are not comfortable wearing the attire of our disorders. The rejection to us for us will provide us a shock, and make us immerse in disbelief. No matter how much we can 'deny' for our despondent conditions, it's inevitable to avoid. We absorb in unwanted guilt not realizing that it's not the remorse that needs to acknowledged rather learning from the undesirable deeds would indemnify better, further saving us from the anger and bargaining which could have followed if chosen the rued road. Depression, loneliness is just a way of life complimenting us sarcastically when we surrender to search or even interpret the solutions. Nevertheless, resentment arises, when we give up the sea to swim in the river, with reconstruction and working through we can own something bigger and better, to happily suffer. With the six stages of grief, which we all face with disbelief, I wonder sometimes why am I not the person I desire to be? Why, in real am not the best "me"?, I have exposed myself to the world I have created, often I am caught in my own "illusory truth", I become downhearted, lately it's been happening, my mind not accepting what is true, rather wanting me to get more bruised. We are not what we pretend to be, unfortunately, our cravings for the ideal "us" but provide toxicity, how do we remove our mask?

Reconstruction-III

Acceptance

My identity isn't any discourse, my identity isn't a debate.

The false parentheses that punctuate my genuine smile often, the sudden urge to isolate sometimes suddenly when people give the most unexpected experiences. Though "it's life" as we say it. Like I am 17, sooner I will be a young adult and gradually will be old enough. Maturity comes with experiences and not with age, which is accepted. But sometimes what I, actually what most of us fail to accept that things can go way up and can go way too deep down. The mask that I have always put on needs to tear down because I felt through all of the years that the more you hide the more it will show, and when it comes to you being 'you', we hardly remember who we are in general and true definitions. The sadness which hits us momentarily, what I believe that life simply expects us to acquiescently accept. For me, my reconstruction and working through the troubles could only come progressively when we are in the know of that everything happens for your reason and this perspective is hard to neglect. Yet we all will face troubles, and most of our questions will be left deliberately unanswered. The era hasn't gone that far, the grieves of ours might leave scars, but with passive acceptance, we can change the scars into marks. My identity isn't any discourse, nor any debate. At some point, everybody gets tired of nebulous turns and wants to be straight. The mask I wear, technically like a fool made me run around. Made me chase something which isn't long term, forget genuine, it's only pretty "outside" vision. A reality I used to live which itself can't figure, how to equally devote and not blabber excuses of rehearsed, repetitive quotes. I have a certain range. Not easy for my heart to bargain with the brain. After making me fall hard, a little more than a few times. I will win with the mask and lose eventually but I will win without the mask and learn gradually. The mask that we wear, isn't just going to fall every time we fall. Unless we learn to accept who we are and little expect from everyone around us. Unless we learn to crave for the mask to fall, suffocate ourselves under the weight of it. We won't want to change. The acceptance is hard in our community but we have to learn to be comfortable in our skin. The day I realized and perceived who I was truly, I felt the soul, when it said: "that I got lost". I noticed the fading roses, to arouse faith and reconstruct myself, the mask had to be destroyed. I am happy that through the stages of sorrows, I accepted them actively and behind the mask, I could be proud to embrace who I was and will be always.

Depression

– Darshini Shah

Depression-I

Present-1

Twenty-five. Is that a number to flaunt? Well, I don't know. But yeah, I am 25 now. Too old right? I am feeling too old. And still, I love hearing stories. Just like a small kid. I love hearing stories about people their lives, their hardships, their struggles, their success, their failures, their happiness, their sadness, everything about them. Even about their horrors, fears and tears. But today I will tell you my story. Story of my hardships and my struggles my happiness and my sadness, my fears and my tears. Mine is a horror story. No, not that horror which we see in movies. Something much much more fearful and scary. My depression story. D-E-P-R-E-S-S-I-O-N.

Depression is not as easy as it seems. A state where you start feeling like everything good has ended in this world and there will be no happiness in the world and no one will ever like you is the state of depression and depression is really not nice. The world itself is tremendously torturing. It is a serious and very common medical illness that negatively affects the way you fell, the way you act and the way you think. It leads to a loss of interest in activities that were once enjoyed thoroughly and that were once found exciting. It can lead to a variety of other emotional and physical problems and can decrease a person's ability to function. There will be changes in appetite and loss or gain in weight which is completely unrelated to dieting. A person can have trouble while sleeping or will sleep too much. The depressed individual may have increased fatigue or will lose energy too quickly. The individual may start feeling worthless or guilty for every small thing. The person may have difficulty in thinking, concentrating and making decisions, and he or she may have thoughts of death and suicide.

I went into depression when I was a teen. Well that is obvious now. Between fifteen and eighteen years of age. That was the worst period, the most dreadful phase of my life. And today, when I am out of that phase, and I look back, I feel that I was so vulnerable at that time. Such inconsequential matters had the power and the capability to drag me into depression. Such petty matters had the potential to affect me so much and so adversely. There were several reasons which paved a path for me towards depression. But there were the principal three events that led me to depression. Read on to find out about them:)

Past-1

Okay, so, Eva, Kaitlin, Lucy, Grace and I pass out from our ninth grade with flying colours. We have had fun in the ninth, perhaps all the fun we could have till eternity. We went out for dinner dates together. We have had harry potter marathons. We have hosted organisation parties where we help in cleaning each other's closets and room and then have a great time gossiping. We made each other's hair and tried new hairstyles on each other. We even enjoy hanging on monkey bars when we are together. We have watched one of the web series straight for forty-eight hours. We have applied nail polishes and make up on each other. We own several matching clothes. We made each other's playlists and complete projects. We have taken countless selfies and posted them on every social media site that has ever existed. We have gone bungee jumping together. That was seriously a nightmare! We have gone for movies, stayed up for night outs, fought for each other and fought with each other. Yet all this has just made us stronger. We had grown a lot more closer during that year. We had become each other's heart and soul. They were my forever, they were my always and they were my constants. We had practically lived and survived together. Their friendship offered me something totally different, which was one of its kinds. They reminded me who I was, why I was special and what was my worth. They never turned back when I was in problems. They made sure I wasn't alone. They knew what were my foibles, my flaws and my short comings. All my disabilities were an open secret to them.

But then all this actually became past tense. Why? Because devils had entered our lives. Devils disguised as guys. First I used to think me Eva and Grace were very close. We shared everything with each other. No force in the universe could ever break us apart. I felt we were inseparable. We won't break and we will be united always was what I used to think. But oh, how wrong I was. It took time in understanding that all of this was nothing but a misunderstanding. Here's introducing you to the first guy who ruined my life: Austin. He and Eva were very fond of each other. They both liked each other, but for no evident reason they had kept it hidden, from the world and for themselves. But the world knew, their actions and their behaviour and their blushes and their smiles and their glances made it obvious. But then, they had come together, of which I was left uninformed. Eva must have not even thought of telling it to me. I don't know why. Grace didn't tell me too. I came to know this when everyone knew. The two people who were the half of my world had betrayed me. Betrayal. I hated betrayal then and hate betrayal now. I don't like people who betray me. It stops me from trusting other people. Trusting a person is like giving the person a gun, putting it at your heart and then hoping that the person won't pull the trigger. Yet we can't refrain from trusting people. That's humanly. And then, something sparked in my mind. The question about Kaitlyn was given air by my mind. For a span of time she had been totally out of my mind. I probably didn't even think about her. How could I do that? I realised this was karma. Kaitlyn was ignored by me and I was ignored by Eva and Grace. I didn't know what and has was she doing. I actually didn't know anything about her teen self. I realize I didn't have a single duo pic with her. There was a flush of guilt in me. I knew it was my fault. But I also knew that she would trust me again, obviously, as she was unknown to my "trust and trigger theory". I went to check on Kaitlyn, how she was doing and what was she doing. And then it dawned on me that she had found herself a new best friend, Lucy. Yet she was friendly with me and helped me a lot. I started liking her. A lot.

I like this guy, his name was Nick. I had an enormous crush on him. He was cute. He was lively. He made the atmosphere lively. His laugh made me laugh. His smile made me smile. His cry made me cry. His frown made me frown. His anger made me angry. His sadness made me sad. His eyes had a different kind of magic which was seldom understood by someone. He had a scar over his left eyebrow, maybe some childhood injury. When he smiled, it was like the universe was smiling. And well, he was the hero of the class. He had a manly stature which made him the crush of half of the girls of the class. No more than half. And you know that was not even a major problem. The even bigger problem was that he liked Kaitlyn, who seemed like she was unaware of even his existence, his presence. She scored more marks than me, she was smarter than me, she knew how to say sassy things and shut anyone up who came in her way, and her prominent cheek bones and collarbones made her look prettier than me. And you know what was even a bigger problem; she was the one who helped me the most out of all my friends. Fortunately unfortunately, she was my best friend. I considered my boon companion, my bosom friend.

This was not the only problem I was facing. As we entered the tenth grade, I was flooded with pressure, stress and expectations. It was the board year, "the tenth grade". Level of competition had crossed all limits and reached all height. People who touched books just before the day of exams had started studying from the start of the year the teachers seemed zombies to me, saying, "Study! It's the tenth grade! The board year!" I had started having nightmares and sleepless nights. My eyes had become all puffy and red and I had become all the more grumpy. I had started scoring lower. My want for Nick gets maximised and I had started hating, literally hating, Kaitlyn. There were continuous tests in the school and in the coaching classes and I was not able to handle this huge amount of pressure on me. I was sad and I felt helpless. My performance degraded and my results had experienced a major downfall. Books had taken a backseat in my priority list. All this was becoming too much for me. I started getting angry on every other person I meet. Small things had attained the ability to irritate me. Everyone had started getting on my nerves. It was as if everyone was trying my patience. Almost everyone raised my hackles. It was people couldn't help but get under my skin and my hair. The mere existence of some people had started driving me crazy. I started getting frustrated on matters which were not even worthy of my consideration.

As if all this was not enough, I got another major blow. Nick proposed Kaitlyn!! A proper knew down proposal with a red rose. Kaitlin seemed flattered. And guess what, she said yes! I was devastated and shattered. I was broke inside. I didn't know what to do.

Present-1

In this manner, Nick became the second guy who ruined my life. Well, to be more precise, he and Austin took me one step towards depression.

Present-2

Teachers. Teachers are supposed to teach us. But they do a lot more than teaching. They turn their students into their own kids and then teach. They get transformed into a mother for every child and then teach. They change the classroom into the home of every child and they teach. They try and give everything to every student and hope for the best for the kid. But sometimes the teachers say or do something which affects a child too much. Especially a teen. One such teacher who had changed me was Mrs. Gill. I still feel the best punishment that could be ever given to someone was to sit in the class of Mrs. Gill. The incident may seem too trivial but at that time I was hurt by what she did. Actually, by what she said.

Past-2

It was the geography class. The old Mrs. Gill was walking was towards the class. I always found her classes to be too boring. I found her too boring. She was in her late seventies and she had no zeal left for life. She seemed to have forgotten what it meant to chill and have fun in life. It appeared that all she had to do in life now was to torcher students like me. Everyone literally hated her class because they were too monotonous, lifeless, soulless, passionless and spiritless. She herself was too uninteresting and unstimulated. She knows how to stultify us, how to stupefy us and how to put us to sleep. Her class are too uneventful as everyone is hardly awake. The chill of the breeze and the mellowness and the comfort of the sun beams coming in the class from the window did not do anything to increase my attentiveness. Instead it seemed they were calling me outside to lie in the half wet and half dry grass and admire what a marvellous it was. But this marvellous day was now going to be wreaked havoc on. Mrs. Gill started droning on and on about latitudes and longitudes and the movement of earth around the sun and grid references and what not. I tend to doze off in her class every time and she had insulted me probably zillions of times. But I just didn't care, firstly because I was not bothered and secondly I couldn't help but sleep in her class. It was getting more and more arduous to stop my eyes from drooping off. I was taking notes as best I could when instantaneously I realized my eyes were closed. I opened my eyes swiftly and glanced at my notebook. The last line I had written on the page ended in nonsensical words after which it became a straight line which ran out of the page. I tried to focus on what Mrs. Gill was saying, only to realize that she was now talking about amount of water on earth and the wildlife variations in the world and then I became indifferent and oblivious of what she was saying.

A few minutes later, I again woke up from a startled doze. I took a quick gander around the class to see if any of my class mates or Mrs. Gill had noticed. A girl gave me about twenty eye rolls in just about two seconds, to tell me she was just as bored as I was. Yet, she was lucky; she was able to stop her eyes from drooping off.

Mrs. Gill had caught me sleeping. Once again. How do I know this? When she began her question answer session, the first student she chose to ask a question was me. I was damn sure I didn't know the answer to whatever she was going to ask. Because I wasn't paying attention. I also knew I was going to be humiliated in front of the entire class. Again. I stoop up. She asked some question and without even listening to the question I told her that I didn't know the answer. She told me I was stupid, I was a disappointment to my parents, I was wasting their money, I didn't deserve to be in this school and then I couldn't care less but hear what she was speaking. Actually I was surprised that I really did hear what she was saying. And then she said something which got all my ears, got all my attention. She told me I was no match to the other students of the class. She had underestimated me. Me! She appeared to be unaware of my abilities in spite of knowing that I was one of the highest scorers of the class. She had wound my self confidence, my self esteem, myself regard, my faith in me, my pride in my abilities. She had rated me too low and I was undervalued. I was brushed aside. I realized how it felt to be glossed over. No one ever had trivialized me so much and so badly. I was aghasted on being belittled by someone who was just an old lady, a boring geography teacher. I was stunned. My self conceit was crushed. My armour proper was demolished and razed to the ground. My inner self was overwhelmed and traumatised. I couldn't even register that I had fallen so low and my image had hit rock bottom. My impression had come to the lowest point ever. I was so gloomy and miserable. Mrs. Gill had made me like herself and her classes, lifeless, spiritless, soulless and passionless. I was crestfallen. I was so heavy hearted. I had become too moody and dolorous. I was in a state of despondency. I had become too pessimistic. I was heartsick and I was totally disconsolate.

Present-2

At that time, Mrs. Gill must have hoped that by saying this I would change myself and focus more on studies. But her words had a completely different effect on me. Little did I know, I was moving closer to the D-word. Don't wanna say it. It still gives me shivers and trembles.

Present-3

I once met a woman who wasn't worried about her weight, her body shape and body image. You know I am joking. There are ladies who really don't care about their body, but none of the ladies I knew had a carefree attitude about their body shape. These "extremely high priority matters" became those of high priority when I was a teen. Well, that was not me only. All the boys and girls who 'teened' with me had the same condition. You too must have gone through something like this too. Or going through something like this. And that's why you are actually reading this. Probably. The surprising thing is, even though everyone goes through the same stage, not everyone goes into the D-state. Some girls and boys who were chubby as kids become slim and slender as teens. But some kids, especially girls, do not change, as far as their body is concerned. I was one of them. Then these girls go out of their way to lose their weight and get into shape. Their emerging and prominent sex hormones give them a crush who won't like them back if they aren't slim and thin and pretty. It was like if you aren't in shape and not good looking you has nothing to do in this world and the world is better off without you. See, issues which seem such trivial as adults today, were like a life and death question when we were teens. I was chubby and overweight from the start. But at that time I was a teddy bear for everyone. And then, when I am a teen, I am fatty and fatso and every other word which was a comment on my body. Now, as an adult, I never think this would take me into depression. Wanna know how? Read on.

Past-3

For as long as I can remember, I have always been somewhat chubby and on the short side for my age. And when I was younger, I dint think much about this and I didn't fuss much about it. I would have fussed about it when others would have noticed my extra pounds or would have thought about it. But I was a cute teddy bear at that time. And I didn't complain. I was enjoying all the attention I got. Everything was just normal and nobody thought I was "the fat girl". But as I got older, I noticed my friends were getting a lot more thinner and a lot more taller. And I wasn't. And when everybody was getting thinner, they obviously noticed that I wasn't. And that's when things began to go downhill. Things got so bad, that too so fast. Doing even small jobs like walking down a street had become difficult. People shouted names right on my face. They gave me strange and weird looks that made me feel so uncomfortable and awkward that I had started feeling ashamed of my own body. A thought came in my mind. Maybe I should tell my mom about this. But then I bashed the thought out of my mind because she already has many problems of her own and I shouldn't disturb her with matters of me. I surfed the internet for diets that would actually work. I started to follow these diets with extreme seriousness and too strictly. But then again, the next day, someone called me another stupid name and gave me another ugly and awkward look which gave me infinite frustration and the only solution I found was chips, coke, pizzas, cheese, burger and all other junk foods. I found solace in them. And then I break my "do or die" diet and start eating normally again. It's not that my mom didn't notice. She asked me several time about what was wrong with me. Sometimes lovingly, and sometimes by the means of fear. But I had become so reserved and so into myself that I always denied whenever she asked me something about the matter. She knew something was not correct. And I wasn't telling her. I don't know how bad she must have felt. Or she must have felt that her child has now grown up and could solve her problems on her own. I don't know.....never been a mom. Well, I am not sure about the problem solving part. I had not yet solved my problem. Every time someone called me a name or passed a comment on my body, I got so annoyed I got out of control. I couldn't think straight and I couldn't comprehend what I was doing and what I was going to do. I couldn't visualise my further actions. I had lost control over my own freaking self.

Autumn and winters had become my favourite seasons, obviously because we had to cover ourselves entirely because of the weather. I couldn't wear all the stylish clothes like the other girls of my age. In any of the parties where all the girls are wearing crop tops, frocks, miniskirts, ripped and skinny jeans and shorts, I was there in a full sleeve tee and a baggy jeans. Eva, Kaitlin, Lucy and Grace tried to help a lot but I had decided a long ago that now I don't want anything from them they were one of the main reasons of my hopeless state. I now had nothing to do with any of them. I didn't even go swimming now, in spite of swimming being my favourite sport. I missed that pools and those dives and those swimsuits.

I tried exercising too. Nobody was going to allow me to go to gym, I was sure about that and so I didn't even ask. I lied about going to tuitions at home and went for jogging. I tried doing push-up and I learnt skipping. I started doing yoga. But I was expecting changes in me a little too fast. I didn't get results and I was so annoyed with everything and I was feeling so hopeless and so sad. I had started hating myself and my body. I was feeling like I had nobody in this world who would understand me and put them in my shoes ever. I felt I would never be able to remove the tag of the "fat girl" that has held me so tightly. I felt I will never be accepted. I felt I had started emitting bad vibes and now no one liked to be me. Well, who likes to be with a sick and helpless and extremely unpopular girl? I had stopped expecting anything from my friends. I had learned long back that no one could solve your problems and you had to face them and solve them yourself. But till that date, I had done nothing but run away from my problems. But I couldn't run away from my own body! I was reaching nowhere in finding solutions towards the problem.

Present-3

I was extremely clueless at that time. I felt I would never get into shape and nobody will ever like me. I felt that things wold never turn out good for me. And when you feel that nothing will ever get solved in your life and you will never be happy, you know which state is that right?

Present-final

And so I had gone in depression. I had stopped eating and I had sleepless nights from day one. It was like the tunnel was never ending, completely filled with darkness, without any spark of happiness and hope. During breaks when everyone was playing around, I used to be that girl who was engrossed in her own world and would be in that corner. I had become that girl who would be crying and probably no one cared. I was never like this. I was always a happy soul. In the first battle between me and depression, depression had emerged victorious. It had successfully taken a toll over me. I had lost too badly. And you know, I don't like losing........

Depression-II

Present-1

I am not getting too boring right? You know I have forgotten what teens find appealing and interesting and what teens find too monotonous and dull. But I will try and make sure that you don't find my depression story too depressing. Oh well, I created an oxymoron there, right? So let me narrate you about the second phase of the battle between me and depression. I told you I didn't like losing. But, they're there, I won't give you any spoilers. You know, I don't mind spoiling your Avengers Endgame, but my own story; no I will not spoil it. There were several incidents which made it evident to my parents that I was in depression. Starting with the first one:

This chapter brings two new characters in my depression story. Now, you must be thinking, "Weren't there enough characters already to make this story too chaotic and tangled to understand?" Well, let me clear this up. All the characters other than me were the friends, who proved that there was no one else who was shittier than them, Mrs. Gill, who proved to be way more wounding then she appeared and guys, who proved that outer beauty was all the wanted. But now, the characters which I am introducing are mandatory and of utmost importance. The two people whom I had hoped to make proud. The two people who had given their everything to me. The two people who had done so much for me. The two people who had sacrificed everything for me, including their passions and likes. The two people who knew what was best for me and made sure that they do that thing for me. The two people who will never consider me as a grown-up and will continue pointing out my mistakes. The two people who will always help me improve myself. The two people who always make me discover more of me, bit by bit, whenever I talk to them. The two people who have taught me to see all the difficulties as small puddles and made me believe that I could easily jump over them and they will be finished. The two people who know what is important to me and what matters to me the most. Yes, my parents. They first shared their DNA with me and now, they share everything me. Their happiness, their sadness, their losses, their profits, their achievements, their money, their hopes, their passions along with their dreams. It's not easy giving up your everything for someone who has done nothing for you, at least for the time being. But they did it. They will always have the highest esteem in my eyes. They will forever be the most celebrated, the most successful, the most lively and the most beautiful parents for me.

Every child is unique and different. Every child has spectacular individual abilities that define that particular child. Every child can achieve his or her personal best when placed on the right path. Every child has a personality. Each child has different views of each topic, even if it is as small as was food particle the ant was carrying when it was in front of them. Their thinking styles are different, their ideologies are different, what ice cream flavor they want or which is their favorite Lindt chocolate also varies from child to child. How does a child complete a task also is different. Every child tries to endeavor and focus and to give his or her best in life. Every child is an expert in one or the other field. Just wait till he or she finds about it and see how that child soars in the wide and open sky. And teens? A teen is just a giant form of a child. Their height increases, their weight increases, their way of speaking changes, their face also shows some alterations. They undergo all the normal hormonal changes which make their thinking kind of mature when compared to a child. Their thinking is too mature for a child and too immature for an adult. Their mindset is still tender and is not capable of withstanding much amount of pressure and stress. The roots of their mindset are still weak, and that is just because a teen has still not encountered experiences where he has to compulsorily undergo enormous amounts of pressure and stress. Just like a child.

Today, when the competition has spread its roots in every walk of life, parents want to see their child just at the top and nowhere else, not even second. And these parents can't help but compare their child to other kids of the same age. They feel that this will help their child to come up to a higher level. But comparing a child at every step just does not help.

Well, "comparison" is counter-productive for every individual. But it is way more demeaning and way more devaluing for children, including teens. Well, especially teens. Why? Because teens think way more maturely for a child and way more immaturely for an adult. If a teen thought as an adult, a comparison would go too deep in a teen's mind. An adult knows what he should let affect him and what he should not let affect him. And if a teen thought like a child, the comparison would only affect his mindset but not his behavior, because children are too innocent. Children, and one way or the other, the teens, are tender beings. Can they take negative criticism positively? No, they cannot. And if this criticism shows a child, let's be more specific there, a child or a teen, how others are better than him or her, it is a lot more painful. But then, this does not mean that parents must not point out a child's mistakes. If we don't show a child his mistakes the scope of improving comes to an end and the child may start feeling that he or she is perfect and there is no one better than them. The child may also start feeling like there is no one around who is worthy of competing with him and he deserves nothing but the best. And if this mindset is not changed quickly, this mindset will get fixed in the mind of the child. And when the child will realize that "Oh! How wrong was I!", the child may be tempted to take extreme," extreme" measures. (I don't need to elaborate on those "extreme" measure, do I?) We don't want that, do we?

If we are told by someone that we are not good at something and that there others out there who are a lot better at this than we are, slowly, yet steadily, mind that, very steadily, we may start doubting our self. A child may also start feeling that he or she can never be better at it. If a child is constantly being compared with other children of the same age, jealousy will surely torment that child. This jealousy can easily take many other divergent forms of negative feelings or emotions like hatred. These feelings of hatred and jealousy can also come out in the form of aggression, fierce aggression.

But when you compare a child, or a teen, in depression, you have created hell for that individual. No something even worse than hell. That's the worst punishment you can ever give to someone, especially a child, and even more especially (there's nothing in the English grammar such as "even more especially", obviously, but just to lay even more emphasis, so...) a teen. There are already thousands of things going inside the head of that individual. Well, the only difference between a depressed teen and a normal teen is that the normal teen does all the daily activities, and depressed teen does all the same activities, just with a lot of extra and added problems. The person is having sleeping problems, eating problems thinking problems crying problems. Comparison arouses the feelings of sadness and hopelessness, which in turn makes it unbearable for that tender being. To be honest, I have had firsthand experience of all of this. So you can trust me ;)

Past-1

My prelim results had been declared and they were not pretty for a student like me. My parents were disheartened. My parents always had a presupposition in their mind that I can never score too low. And today, when I had scored too low, obviously, they were dispirited. They were feeling baffled and frustrated with me. I can't see them being sad. They were so precious to me that I can't see them hurt. But more than being hurt and sad, they seemed so angry and so irate and so vexed and so displeased. I felt like I was going to be cast down in the Tartarus and I was going to face eternal damnation and eternal punishment. I was feeling like there was no other stare that was crueler than my mom's stare and there was no other slap which hurt more than my dad's slap. And because I think that you are thinking that I am exaggerating a lot, let me tell you I am not because I was in depression and everything automatically gets exaggerated. There was already too much distress in my life and it was beyond my control and out of my hands to handle any more of this 'depression type' emotions. I was a deterrent, a complication, a snag in my path. I was in no need of any more of those hindrances. This is because even if hindrances came in my way, I was not in a position to remove them from my way. I was in a state where I was like "Oh hi Hindrance! Do you think your pals were not enough that you have also come? Oh but never mind, I am sitting here, you also sit here. If you think you want to go, then go, otherwise, you try and enjoy here, just like your pals."

When I had come home, my mom gave me a glass of water. She made me sit down. She said, "It is okay if you don't score well." But deep down, we both knew how much marks mattered to my mom and it was not, absolutely not okay for her if I didn't score well. She then continued, "But your marks prove that this time you didn't work hard. You know you cannot even think of success without working hard. To succeed in life and to excel in life, you must be committed to your job, you must be conscientious and you must be diligent. When you can see that others are so ahead of you why can't you start working hard for yourself and why do you wait till I scold you like this? And look at your score in the last exam and look at your score in this exam? Instead of increasing, it has gone so low! Now is this what you are going to show us after all the things we do for you and all the things we have given you? You should at least aim to score more than what you scored in the last exam. We can at least expect this much from you, can we? Look at Kaitlyn, she got eighty-five percent in the terminals, and now she has got ninety-two percent in this prelim. And look at you; you got ninety-four percent in the terminals and now, an eighty-three percent. Is this even acceptable now? If you don't score now, you know what will people think about your ninety-four percent in the terminals?" She was walking from this side to the other. It was so difficult to concentrate on where she was walking and what she was speaking. Looking at her walk and listening to what she was speaking was making my mind throb like anything. "People will say it was just a kind of a beginner's luck because you have never scored so high. And you know what kind of impression this will create about you, right? Your teacher also praised you so much. And what is the result of all of this? Isn't this too bad about you? You have to do a lot of hard work now." She had turned around now. She continued, "Look at Eva;

she is also working hard now. Look at her marks seventy-nine isn't too bad for people at her level. And Aron did well too; eighty-two percent is not too bad for him, is it? And look at you; you couldn't even get a nine in the front. And you and Janet were at the same level right now... Janet has scored exceptionally well....what has happened to you? Well, look at Christina and Andrew; they are the toppers of your class right? When you all were young, they both were a lot more behind than you. And you didn't even realize when they left you way behind. Long gone are the days when I thought you would become the school topper. You better buck up now. You don't forget that you were going to prove yourself. Now why don't you speak something?!" her voice was loud I jumped on my place. But my mom didn't see this because she had her back turned towards my face.

I started speaking. "I know mom. I know. But I think it's too late to overtake them now mom. I don't think I can do it. I don't even deserve good marks because I don't work hard for them. It is all my fault mom. I should have studied a lot harder mom. Please forgive me mom.... mom, please... I am so sorry mom. I don't think I can be better than Kaitlyn or Andrew or Christina, maybe even Janet. They have become a lot smarter than me now. I am no match for all of the mom. They are all way ahead of me. I am sorry mom. It is all my bloody damn fault mom. Mom....." before I could say any further, I was breathing too fast, I was vigorously hiccupping. Tears were rolling down my cheeks as rain falls on earth. I was going to throw the nearest flower pot on the ground. You won't believe me but I don't remember standing up and taking the flower pot in my hands. My mom had caught the pot and that was when I realized what I was doing.

Present-1

My mom looked horrified at that time. She knew that I was a person of self-praise. She knew I never accepted my mistakes. I didn't like myself being proven wrong. I was full of boastfulness about myself at that time. I never showed modesty. So...well...

Present-2

People who sit with you, stand with you, eat with you, sleep with you, joke with you, cry with you, laugh with you, pose for pictures with you, travel the world with you, will undoubtedly notice when there are changes in you, above all, changes in your behaviour and your habits. When my tenth grade had just started, I was too excited and determined to score good marks and prove it to everyone that I was not the one you can take for granted. But when the time for board exams came, I was already in depression and there were a lot of things going on in my head. It had become really difficult for me to concentrate on one thing at a time. Everything seemed different and changed around me and in me. I started feeling clueless about each and everything.

Past-2

Board exams were round the corner and high tension had settled around the atmosphere just like high pressure settles down. My prelim results were out and my scores were so substandard that they could be appreciated. My parents had much higher expectations of me. And they had a right to have high expectations. I was a bright child. I used to keep books in front of me all day and my parents thought I was going to score the highest and there was nobody who could beat me. They had extra faith in me and told "Our daughter has it in her, she was brilliant in almost everything. Her potential level is higher than everybody in the school who studies in her grade," to everybody they met. And mind you, I was brilliant. But then, I was going through a major crisis in my head which was nowhere near getting solved. After my rock bottom prelim result, my parents decided to make me study together because they could not bear seeing me falling so behind. And because I had become numb to all the slaps and admonishments I got from my parents. For the first few days, they sat with me and made me learn and write. But then, after a few more days, things got back to what was the usual and I got back to square one. Every day my parents used to ask me, "How much of the syllabus have you covered today?" I used to say the names of some random chapters and then they would feel happy and me, I would get satisfied with the lie that I used to speak every day. At least my parents were getting happy. They felt that I had mended my ways and was now moving on the right track. They felt that I had again got the 'will do anything to score good marks' attitude. And me, I was anesthetic to all of this too. Reality punched in the face of my parents when I came home holding my board result. Eighty. Just an eight in the front and zero at the back. My parents were left aghast. There were tears in the eyes of my mom. My eyes were not dry too.

I had never seen my parents so helpless and so sad and so broke and so inconsolable and so grief-stricken and so miserable and so dejected. This is what will always have the ability to soften me up and melt me down. I started crying harder and now, it was their turn to soften up and meltdown. They told me, "Learn to accept things. You did less amount of hard work and so you got fewer amounts of marks. Stop crying over spilled milk now. Be strong and think about how you are going to tackle your future. If you keep crying like this now, you will keep crying like this in the coming future too. You have lost just one opportunity to score good marks. We know there are many more exciting opportunities in store for you. Get up and start working right now. We hope that this has taught you an important lesson in life. And we expect you to prove that you have learned something."

I had always been too passionate and too perfervid when it came to my studies. I never did anything which would adversely affect my studies. I would put my everything in books when I was studying and it was difficult to shift my focus to anything when I was learning and writing. My mom knew this. She knew each bit of me very well. When I had scored too low in my prelims, she thought I will work very hard for my board exams. But then I didn't score well in my boards too. And so my parents knew there was something wrong with me and in me. But they didn't speak anything about this to me. They started talking a lot more politely with me. They had started doing everything I wanted and gave me everything I asked for. Every night they used to come and sit every night and talk to me about how was my day and all that kind of stuff. They had stopped giving me outside food. They had started giving me homemade food all the time, not even chips and coke. But I wasn't concerned about what I was eating. All I wanted was that I should be eating something or the other. They had changed their attitude towards me, thinking that this would help. They now had stopped boasting. People threw taunts at them but they didn't give a damn to those taunts. All they cared about was me and my future.

Present-2

I had given so much pain to my parents. My parents had got this in return for their enormous contribution to my life. That day, I had realized how beautiful my parents were. That day I had also realized that they know me too well and I should have told them there and then about what was wrong with me. They have been wonderful parents for me.

Present-3

Anxiety. It is the intense excessive and persistent worry and fear about everyday situations. Fast heart rate, rapid breathing, sweating and feeling tired may occur. It is only an indicator of underlying disease when feelings become excessive. It is a normal emotion that causes increased alertness, fears and physical signs such as a rapid heart rate. Anxiety is more than just feeling stressed or worried. Anxiety is when these anxious feelings don't go away, when they are ongoing and happen without any particular reason or cause. Researchers don't know what exactly brings on anxiety disorders. Like other forms of mental illness, they stem from a combination of things including changes in your brain, environmental stress or even your genes. This disorder can run in families and would be linked to faulty circuits in the brain that controls fear and other emotions. Well, all of this is just medical nonsense. The chief thing that you should remember about anxiety to continue reading is that ANXIETY IS THE BIGGEST SYMPTOM OF DEPRESSION. And yes I am shouting. Well, that had to be related to depression. I just hope you weren't expecting something else. And if you were, sorry I destroyed your hopes :P

Past-3

I have been a calm decision-maker from the start. I never got over-excited over any crucial decisions that I had to ever make. But well, is there anything normal when you are in a depression? No, it isn't. Nothing remains the same. Not a single thing. It feels like you have landed up in a completely different planet and people around start seeming like aliens. You just feel so out of place. Even places like the open ground with the seemingly wet grass start feeling claustrophobic. You become so engrossed in your own life that you start to shut your eyes and turn a deaf ear to all the other people. Okay, so whenever I had to make an important decision, I would lower my voice, calm down and would focus on not making any sudden moves. I would stay in control to try and not make any rational decisions. I neither get too high nor too low when I am confronted with a difficult decision. I try to visualize, in advance, the outcome of a decision. I always try and consider what will be problematic and what will be beneficial for me. My parents considered this as one of my unparallel and unmatched qualities. Many a time they come and ask me to help them in making their own decisions. They even told me "You should publish a guide to decision making. You could earn a lot you know." I would just take it as a joke and put that matter out of my mind. But this became the greatest clue for my parents. My this quality confirmed their suspicions, exactly as they had felt, not expected. And as their suspicions got confirmed one by one, they got scared more and more. My this quality allowed my parents, especially my mom, to delve deeper in my mind. They had started formulating stories in their head about what must have happened to me. I had no idea about all of this because I was way too busy fighting my war. I was so immersed in my self.

For most students in grade nine and ten, visions of a career path are mostly restricted to childhood dreams like "I will grow up to be a doctor" or "I will grow to be an engineer" or "I will grow up to be a pilot", or "I will grow up to be a teacher", or........ Well, you get the idea right? Ever since of those early flights of thoughts and dreams that aspired to be completed had taken off, very few of these get completed. Yet as ninth grade ends or tenth grade begins, all of a sudden, these questions start coming up in a big way, a way which those children who had dreamt to become teachers and doctors and engineers and pilots had never expected. It is like a time bomb for the ninth or tenth graders, you answer or the time bomb goes off. And although the time bomb may not be successful in taking your life at present, it is surely going to take it away in the future. How? It is simple. When you are not ready with an answer in time, you will surely end up making the wrong choice. When you make the wrong choice you will end up in a career which you don't like. And when you end up in a career you don't like, isn't your life totally over? Parents, teachers, friends and all the people around can't stop talking about this all of a sudden. And when all these years of growing up you have just laughed this question away, how could you answer it, all of a sudden?

Everybody keeps on preaching about doing what you love and do what you want, but no, it isn't that easy. Everyone just preaches that to be happy in life, you must do the work you love. But there is nobody to show the way to reach that point where a student imagines his or her life. And don't you think that a fifteen-year-old child is too young to say what he or she wants from life?

I have been in the Indian Certificate of Secondary Education (ICSE) board from the start. When a student passes the tenth grade, the ICSE board offers three streams in the eleventh grade: Science, Commerce, and Humanities. An eleventh grader must choose one of these streams. The stream that an eleventh grader chooses has a direct connection to his future career options of that student. The stream that a student chooses will go a long way in deciding that child's future. The stream that the eleventh grader chooses will decide whether that student will live his life happily or whether he or she will not live at all. This stream determines how much progress a person can make and how much a person can be successful. This stream decides what kind of life that individual will have in the future. Too much power in just one choice right?

To choose science, you must be a science freak. To choose commerce, you must be a commerce devotee. And to choose humanities, you must be a humanities fanatic. And I was none. My abilities demanded science, my mind demanded commerce and my heart demanded humanities. I kept on asking every person I met on what stream I should choose. My parents were not very much stressed about this, because they thought they knew that in the end, I was going to make the right choice. But they were ignorant of the confusion and the chaos and the perplexity present in my mind then. I was confused about what should I do. During my sleepless nights, when I used to lay on my bed with my eyes wide open and think about what I should do, I would keep on thinking for hours and hours and still, I would be left without any answer. I used to get very frustrated then I could feel my heart rate increasing. My breath would quicken and there seemed tightness in my throat and chest. I was never able to control the situation. And then, I used to feel frozen, my brain used to trigger a freeze response. I always tried to make anxiety go away by distracting my mind, although I knew that it would never work. It was then I realized that fear was wired with anxiety. At nights, when I would become a victim of the side effects of anxiety, I used to get scared out of my wits and think "What would happen if mom and dad got to know about this? They would be so ashamed of me. If they came to know about this I would take them into depression too. I can't let that happen. No-no-no." I just wonder today that why did I didn't think even once that my parents will help me get out of this fix. I continuously ponder upon why I had to even think that it was embarrassing to be in depression. I used to reflect upon why did I think that all of this mess had to be solved by me and only me, just because all of this was MY mess, I could get help from outer sources too. Only if I had raised the amount of courage that was needed to speak up. Only if I could understand that it was okay, completely okay to be in depression. Only if I could realize that I wasn't actually from a different planet, and others like me were suffering from the same issue. Only if I had to ability to decipher that depression was nothing but normal and it wasn't a big deal. Well, of course, it was a big deal, but at that time if I had thought that it wasn't a big deal, I would have been in a better state that day. But yeah, what is done is done.

And finally came the day when I had to submit my answer. The life I wanted was just a click away. But I still was unsure and undecided of what I wanted. When I was in my right mind, I had planned so many things for myself that I wanted to do this and I wanted to do that. I had so much in my mind. I used to think that my life was all set and I will not have any confusion in the future. All that planning started from this stage. And when I had a chance to take destiny in my hands and shape it as I wanted, I was unable to do anything because I didn't remember a single thing. I was so upset with myself. I was asking the earth to swallow me up. I felt like I was going to be doomed now.

I then tried using my trick. I calmed down. I lowered my voice. I started talking to myself because I needed some expert advice. I was first speaking so slowly that only I could hear me. I said, "I know I can do well in science. And the job aspects of science are very bright too. Plus that the salary some prestigious companies offer is way more than the salary in all other fields. On top of that, the employee benefits you get in almost all companies are not something that can be neglected. But to attain all of these benefits the eleventh and twelfth grades ask for a lot of hard work. But commerce is way easier. You don't have to do much hard work over there. It just needs a bit of logical thinking and then you are done with it. But there are not many options available to graduate from when you are taking commerce. There are just a few selected fields in which you can go. There is no wide range of options. But I wanted to become a journalist when I was small. I can't become a journalist if I choose science or commerce. But you don't earn much if you become a journalist", I was remembering some parts of those plans of mine, everything wasn't gone yet. I continued, "If I make one wrong choice today, my life will go into rack and ruins. If I make a wrong choice today, the victim of that will be my tomorrow. My tomorrow will be finished. If I don't study well in the stream I choose, my career will be destroyed! I won't get a job! And if I won't get a job, I won't have food to eat and house to live in. And if I won't have a house, I will come on the road! I can't let that happen. No, I just won't let that happen! If I come on the road, what will I do then?! Oh no, I will be devastated! There will be no one to help me! I will be jobless! I will be a penny less! I will be homeless! I will be friendless! I will e helpless I will be.....I will be......" and then I was about to faint. That was when I had come back to my senses. I had been speaking so loudly that my neighbors had come to check whether everything was alright. And all along, I was thinking that only I could hear me speaking. I was pulling my hair. I was pinching my self. I was breathing too fast. I was going on crying and going on speaking. People had gathered around me and I hadn't noticed even that. I was so into myself. And when I was about to faint, my mom had caught me, made me sit, and gave me a glass of water. She carefully placed me on my bed, switched on the air conditioner, and put me to sleep. Well, she thought I was sleeping. When I couldn't sleep in the night, how could I sleep in the day? But somewhere inside, at that time too, I knew I wasn't successful in fooling her. But I was scared. I was too terror struck. My sweat felt cold on my skin. All the color from my face had disappeared. I wasn't able to guess what my mom was thinking. I was confident that my mom had already called my dad and he was on his way home from work.

Well, to end your anxiety, I had taken science with maths at that time........ Maybe, that day I should not have started speaking only, I never knew I would get so loud. It had become almost, almost crystal clear that something was unsound and faulty and flawed in me. But everything that was happening was happening for the best. Without my knowledge, my parents had already spoken to a family psychologist about my situation. They were too disturbed by this unnatural and abnormal behavior. I am so unhappy that I had hurt then so much.

They already had too much to worry about, and I was an extra. But they had proved that there was nothing more important to them than me. They had proved that my happiness was all they wanted.

Present-3

I don't know how I remember all these incidents so vividly and so well. My parents came to know why I was sleeping so less and eating so much. The second phase of the battle included not much of me. My parents were the Ironman and the Thor of my story. (although I am making a lot of Avenger-ical references, let me tell you I am not, yes you read that right, I am not a Marvel fan) but yeah, my parents had put me on the winning path. I love them 3000(here I go again...). And yes I don't like losing.

Present-4

Need of a recap right? I feel the same too okay so, well, I go into depression because of my 'free-enemies' guys and Mrs. Gill. They were the ones who made me feel unworthy, unhappy, undeserving, uncomfortable, unsatisfied, unnatural, underprivileged, unchallengeable, underestimated. Then my parents got to know about this unstable condition of mine. They first tried to dismiss these thoughts as foolish and impractical. They seemed to believe that for things like depression and all other serious disorders, I was out of bounds. These things did not apply to me. They also seemed to think that I was not entitled to all of these kinds of sufferings and nothing should ever happen to me. They also felt that they were not able to protect me from all of this. They thought they thought that they have been too careless in my case. That had always given me the things I wanted and asked for. They have always given me the freedom of doing most of the things I wanted. Whenever they asked me some questions related to my friends or something about the school, and I wasn't willing to answer, they were always okay with it and never tried to get things out of my mouth. They thought it was all their error and their fault and their mistake that they had given me so much privacy, thinking that it would bring me closer to them, much closer than I already was. But still, we drifted apart. Like the waves move away from the shore, we had drifted apart just like that. My parents and I were like the two planets of the solar system, we know they are of the same kind and belong to the same sun family, yet their paths around the sun could never be the same… we were just like that… We were like the waters of two different oceans. They have the same chemical formula, yet they don't match with each other. We were just like that, so close, yet so far. And now let's go ahead the last phase. This is my favorite, you know.

Past-4

Psychiatrists. Try and listen to this word in that scary ghost voice. PSY-CHIA-TRISTS. Because that's how I used to hear this word. A psychiatrist is a special type of physician who specializes in psychiatry, a branch of medicine that studies the diagnosis and the prevention of medical disorders. My parents had fixed an appointment with one of the psychiatrists. This was what they had told me. They had fixed an appointment with almost all of the psychiatrists of the city. You know I still have to RY to not get them wrong. I sometimes feel so bad that I can just get it into my head that what they were doing was going to benefit me in the end. I just can't get it through my idiotic brain that all they are doing since my birth is for ME.

Whenever you enter a psychiatrist's room, you start wondering what he will tell you to do. Will he ask you to lie down and check your chest with a stethoscope (well obviously he isn't going to check your chest when you have got problems going on inside your head, but still.), or will he tell you to describe your dreams, your experience, your fears and about why did you shed all those tears. He may ask you about your childhood or your parents or anything else. Sometimes it is this fear of the unknown that keeps those in actual need from getting this help. But all the other times, it is this. People think that all kinds of mental diseases are the same: you are crazy, that's it. This is the major reason why people don't get cured of these mental diseases. When we think of diseases like diabetes or jaundice or dengue or some kind of heart diseases, we don't even wait for even a few seconds to start the treatment. When we are beginning to show those symptoms, we try, almost immediately, to reverse these symptoms. We don't ignore them. We develop a plan of action to try and stop the disease from making any more progress. We don't take mental illness too seriously, mostly because of society. We must lift the veil and see how dangerous these diseases are.

Okay, so after creating so much commotion and getting so many slaps and pleading to not go where I was at that time, I was there. In front of the office of my psychiatrist. I told my parents that I was not crazy; I was all jolly and good. But they had ragged me to this seemingly gloomy place. My legs felt stiff and my mouth behaved as if it was zipped. I went in, half-hearted. I crossed my legs and waited for her to speak. She made green tea for me. The first thing she told me was that all we were going to discuss would be kept confidential and I could tell my friends that I was going to an art class or something... I said "Okay", and I thought "As if they care." And the next thing she said was "You are in depression." She said it in such a way that for a second I was like, okay, so what? He continued, "I have studied your symptoms and it is too obvious that you are in depression." I felt the heat rising from my toes and that heat reached my mouth. That heat melted the zip of my mouth and I shout, ' How many times do I have to tell it to everyone? Hear it once and for all! I AM NOT IN DEPRESSION!!!" She shouted, at the same volume, "ACCEPT IT! IT IS NOT SOMETHING THAT YOU FEEL EMBARRASSED ABOUT!" I sat down quietly. (Not everyone could do that you know... That was pretty impressive). She started explaining, now calmly, that it was okay and that I was not abnormal. She tried and made me feel comfortable. I told her every single thing that happened to me, how I felt at that time, how I reacted to it and what were the consequences later. That fear, that awkwardness, that uncomfortableness, that anxiety, had all evaporated. She had become a very nice friend of mine. She was my about mother's age and a mother too. And then I realized that I didn't even know her name. I left the cabin with a smile on my face, (surprisingly) and read her name written on the door of the cabin. Dr. Isabel Evans. "When you feel that everyone hates you, just think 'that's ridiculous, everyone docent know you yet."

I was impressed with her. That day I had learned that I was in depression. (I knew this already. I had just finally accepted it that day.) And I had learned to accept failures, sad incidents, and negative emotions. I was looking forward to more sessions with her. They seemed fun. I went home and I was kind of happy and now, that was unusual (well of course). And maybe some of it, I had let it be shown on my face. My parents were trying to exchange glances, and I had seen them. Oh, it was good to see them happy again.

And after the eternal and ceaseless wait of 2 days, I was getting ready for my next session. I wore a pretty yellow-colored floral frock, put on lipstick, applied mascara and sprayed perfume on myself. I wore my prettiest and favorite sandals. I reached her office and she was already ready with my treatment plan. Little did she know, I had already started healing? Mrs. Evans said that she thought psychotherapy was the best option for me because my depression was not so serious that I take drugs or medicines for it. Psychotherapy is the treatment of all kinds of mental illnesses by discussing, just discussing problems and other causes of it and the solutions too. This is the first tool that every psychiatrist employs to his patients. But Mrs. Evans was sure that I will be done with this. She has made the plan following our first session where I told her what happened. I tried and peeped in her notepad, and this was written on it:

Session 1: Ask her the cause

Session 2: make the psychotherapy work

Session 3: set her free

Her small but clear plan had an immense effect on me. She had already decided how she was going to 'set me free'. Today in my psychotherapy session she told me that in teenage years, guys start mattering to girls. That was okay. But she asked me, "why did it affect you so much that you start scoring so less and you start judging yourself because of them? Make your mind strong. If your mind is strong and bold you will only be registering things which are important and will matter to you. Those things will give you happiness. Those things will ignite a fire in you. Those things will give voice to your heart. Those things will take you where your satisfaction belongs, where you belong. Those things will give you a goal in life. They will inspire you inside you." I was left astounded. I was stupefied. She then said, "Mrs. Gill must have never thought that her one sentence could affect you so much. She knew you didn't listen to what she said when she scolded you. So she got harsher every time she scolded you, hoping you would listen. Teachers are blessings, true blessings dear. Be thankful for them. Learn to appreciate things in life. Sometimes you get them just once in life." She was truly a role model for me. She had left me open-mouthed. And after listening to many more of those types of philosophical sayings, I was already set free. I was in no need of the third session. I was done with depression.

Depression-III

Acceptance

Acceptance is the act of accepting or being accepted. Acceptance is when you take a situation or a person or any random thing as it is and without making any kind of attempt to change it. acceptance is when you just believe that some things are going to be just the way they are and are never going to change. Your mind just starts thinking that no force in the universe can change it. acceptance is knowing that there is nothing in your power that you can do to change something and it will remain, and for some reason, it must remain as it is, and you are completely helpless about it.

Almost every second we are trying to change things. Whether it is a person, a situation, a condition, we d not care about it. All we are focusing on is how to change things which are unfavourable to us. But what we don't seem to realize is that we cannot change a single thing without first accepting it. you can probably mould the future and give it the shape you want. But the present and the past.... they are de rigueur to accept. We cannot so anything but accept it. just because there is nothing we can do to change it. we know that change is inevitable, but accepting your present situations increases the chances of that change being in your favour and to be what you want them to be.

For me, the word 'acceptance' itself radiated happiness.

> *"Happiness can exist only in acceptance."*
> – George Orwell

In some situations acceptance is the only way to be happy. The agony when it is time for someone you love dearly to go to heaven can only be reduced when you move on, stop waiting for the person to return, and accepting that the person is not going to look back and had now moved to a different world. Resistance to reality creates an incredible amount of stress, pressure on your thoughts, your actions, your speaking skills, your hearing skills, on each of your functioning cell. You waste a tremendous amount of energy in denying yourself from the truth. When you accept it, you have a lot of energy, hopefully positive, at your disposal since you no longer have to convince your heart into believing what actually isn't the truth. When you are fired from your job, just accepting the fact that you are actually fired will turn you into a better problem solver. You place yourself in a better position to consider all the prominent and promising options that you have. Your thinking gets cleared, and you choose an appropriate plan to get out of the mess that you are in. Rejecting reality does not change it. Accepting it makes it easier. When you choose acceptance, you choose happiness.

Yes acceptance is a choice. One of the hardest of choices you will ever get, but a choice at any rate. You solve half of your difficulties by just accepting them. When you accept, you are halfway to hope, you are halfway to the calm, you are halfway to victory, and you are halfway to your goal, your aim. The other half seems to have turned easier too. Acceptance of myself, my problems, my destiny, my mysteries, my strengths, my weaknesses have helped me to resolve, fathom and straighten out the mayhem, the pandemonium, the turmoil that sometimes seem to have permanently established itself inside me. Acceptance clears my thinking. The nucleus of my thoughts is not the problem, it is the way to solve the problem. And now, the problem does not remain much of a problem. It becomes just like the last hurdle of the hurdle race. Just two steps more and victory is waiting for you with open arms. Whne you keep toying with just the first hurdle of the race, that is, not focusing on what is the problem and actually giving your attention to how to solve it, you may probably never know the bliss of winning the race.

Acceptance can be zillion things to zillion people. Acceptance can be in the form of love, compassion, trust, comfort, satisfaction, contentedness, courage, hope, peace, euphoria, even the key to success. Acceptance is probably the greatest of the deeds in the world, the noblest of all. Acceptance makes you ready and turns you strong enough for all the obstacles that you have to cross in your life. When you accept people and situations, they accept you in return too. Accept for peace. Accept for serenity. Accept for yearning. Accept for valour. Accept for staying alive, living a life worth living.

Mentors

Anjana Anand is the co-founder of FYCGlobal Career Guidance Company. The energy of teens fuels her passion of making a difference in their lives. A constant learner herself, Anjana is a certified career counsellor from University of California, Los Angeles, a certified NLP (neuro linguistic programming) practitioner, a communications and social development expert and a trainer at heart. She has over 25 plus years of professional experience of working with children in India, Singapore and Bangladesh. When not mentoring students for careers ahead, she indulges in running marathons, reading and writing.

Her Belief: You project what you feel within. Once you understand your inner self, you can use it to your advantage and make a lasting impact in your career and life.

Yashi Shukla

Yashi Shukla is the co-founder of FYCGLobal Career Guidance Company. She is a certified career counselor from University of California, Los Angeles. Yashi has the innate talent of creating a comfort zone for teens where they are happy to be just themselves. Clubbed with this, she has 15 plus years of experience of working in the education space in India successfully placing students for higher education across the globe. When not mentoring teens, Yashi indulges in yoga and weight training to nurture her passion for physical fitness.

Editor

Anoushka Ray is an IBDP student at TSRS, Gurgaon, Haryana with a passion for literature. Through years of reading works by countless other authors and writing her small pieces, she always dreamt of being a part of the editing process, to help aspiring authors realize their potential- this motivation, along with Anoushka's first hand experience with anxiety, depression and ADD contributed to her desire to be a part of "The Visage: Unmasked"

Co-Authors

Darshini Shah was born and brought up in Valsad, a small and quiet town in Gujarat. She is a grade 10 student at Atul Vidyala. Her poems featured in the school magazine and encouraged her to nurture her passion for writing. Her short story took her to Katha Writer's Workshop in 2015. Books for her act as instruments of inspiration, happiness, hope, and belief for her. Darshini is an orator and an accomplished Bharatnatyam dancer.

Ira Bhattacharjee is a grade 10 student of the Heritage Xperiential Learning School. A passionate writer and an aspiring author, she loves the power of words that create the intoxicating feeling from developing her own world or diving into others'.

Nandini Shukla is an ardent student who loves exploring new horizons. A grade 11 student at Shalom Hills International School, Gurgaon she is a passionate and accomplished Bachata dancer and loves to pen her thoughts.

Navya Sheoran is a grade 12 student at Suncity school, Gurgaon. She loves to express her feelings through her writing

Sadhika Anand is a teenager born and raiosed in New Delhi. She studies in Bhatnagar International School and has always been an avid reader. Writing was an important part of her life growing up as she felt that her pen and paper were the only things that truly understood

her. Sadhika is also the founder of hearNhelp, a non-profit organisation that works to spread mental health awareness.

Samira Bhayana is a grade ten student at TSRS, Gurgaon. For her, mental health awareness is the need of the hour and must be addressed with utmost importance.

www.ingramcontent.com/pod-product-compliance
Lightning Source LLC
Chambersburg PA
CBHW020741180526
45163CB00001B/309